Silent Snow

Silent Snow

The Slow Poisoning
of the Arctic

Marla Cone

Published simultaneously in Canada
Printed in the United States of America

Interior artwork by Steve Gugerty, www.gugerty.com

"Magic Words" from MAGIC WORDS, copyright © 1968, 1967 by Edward Field, reprinted by permission of Harcourt, Inc.

FIRST EDITION

Library of Congress Cataloging-in-Publication Data
Cone, Marla.
    Silent snow : the slow poisoning of the Arctic / Marla Cone.
       p. cm.
    Includes bibliographical references and index.
    ISBN 0-8021-1797-X
    1. Pollution—Arctic regions.   2. Pollution—Environmental aspects—
Arctic regions.   I. Title.
    TD190.5.C66 2005
    363.738'4'09113—dc22   2004060707

Grove Press
an imprint of Grove/Atlantic, Inc.
841 Broadway
New York, NY 10003

05 06 07 08 09   10 9 8 7 6 5 4 3 2 1

For Christopher

*May you find an Inuksuk to guide you, and
may the world you leave behind be better
than the one we left you.*

Courtesy of www.exxun.com

Contents

The most alarming of all man's assaults upon the environment is the contamination of air, earth, rivers, and sea with dangerous and even lethal materials. . . . In this now universal contamination of the environment, chemicals are the sinister and little-recognized partners of radiation in changing the very nature of the world—the very nature of its life.

—Rachel Carson, *Silent Spring*, 1962

This is a land where airplanes track icebergs the size of Cleveland and polar bears fly down out of the stars. It is a region, like the desert, rich with metaphor, with adumbration. In a simple bow from the waist before the nest of the horned lark, you are able to stake your life, again, in what you dream.

—Barry Lopez, *Arctic Dreams*, 1986

Under nature, the slightest differences of structure or constitution may well turn the nicely balanced scale in the struggle for life, and so be preserved. How fleeting are the wishes and efforts of man! how short his time! and consequently how poor will be his results, compared with those accumulated by Nature during whole geological periods! Can we wonder, then, that Nature's productions should be far "truer" in character than man's productions; that they should be infinitely better adapted to the most complex conditions of life, and should plainly bear the stamp of far higher workmanship?

—Charles Darwin, *The Origin of Species*,
Chapter IV: Natural Selection; or the Survival of the Fittest, 1859

The Great Sea has set me
in motion
Set me adrift
And I move as a weed in
the river.
The arch of sky
And mightiness of storms
Encompasses me,
And I am left
Trembling with joy.

—Uvavnuk, of Igloolik, as translated by Tegoodligak,
of South Baffin Island

Introduction

A Moral Compass in a Vast, Lonely Land

Keeping vigil throughout the Arctic, stone sculptures called Inuksuit can stand up to the brutal forces of wind and ice for centuries. Each, like its human creator, is borne of the land and remains part of the land, and its strength lies in its unity—remove one stone and the rest tumble to the ground. As a practical matter, the piles of stones—arranged in the shape of a human being, often with outstretched arms—are compasses that guide lost travelers toward a safe passage home or toward their best hunting grounds. Ethereally, they are moral compasses, icons of the Arctic's human spirit, artistic messengers of trust and connectivity in a vast, harsh, lonely land. If an Inuit hunter sees an Inuksuk, he knows he is not alone. In a place where the sun never rises in winter and never sets in summer, human beings rely on the cooperation of nature, as well as each other, to survive. In Inuit legends, humans and animals are interchangeable. Man and his prey transform into each other, the body and spirit of a human melding with the body and spirit of a seal or bird or ox, seeking the fragile balance that allows each to survive, even thrive, without overcoming the other. In the Arctic, the strongest ties that bind are the ones that bind people to nature. Throughout the centuries, there has been no more sacred and everlasting bond, even between husband and wife or mother and child.

I learned this for myself the day I accompanied narwhal hunters on a sledge trip across the frozen, glacier-carved terrain of Qaanaaq, Greenland. Suddenly, the dogs sprinted off after a ringed seal before I could scramble aboard the sledge. The dogs and hunters vanished into the horizon within seconds, leaving me alone on the ice. Slowly, I began to walk, my boots sinking into the thick snow as I followed their tracks. I knew they would return for me, but momentarily, I felt the desolation and danger of the Arctic as never before. I had no Inuksuk, spiritual or physical, to guide me, and there was no doubt in my mind that if the Inuit hunters did not return, I surely would die, and die quickly. A century ago, hundreds of white polar explorers tried to reach the North Pole, and the only ones who managed to survive were those who abandoned their modern ways and relied on the know-how of Arctic inhabitants. It's the only reason I survived, too.

The Arctic is the last place on Earth I expected to find the world's most severe toxic contamination. I have been a journalist for twenty-six years, covering environmental issues for nineteen of them, writing about all the various ills that human beings have inflicted, often unwittingly, upon this planet. But few have astonished me and intrigued me as much as the discovery of a toxic legacy haunting the Arctic. It was 1996, as I was researching a series of articles for the *Los Angeles Times* about contaminants that suppress immune systems, when I learned of a phenomenon I have come to call the Arctic Paradox. I was questioning experts about where I could find the most heavily exposed communities, where people are harmed the most by toxic pollutants, particularly industrial compounds called polychlorinated biphenyls (PCBs), considered one of the worst environmental villains of the twentieth century. They had been linked to cancer, reproductive effects, and damage to developing brains as well as compromised immune systems. I expected the Great Lakes or the Baltic region, bastions of the industrialized world, to be the hot spots. When I learned the answer—an obscure place called Nunavik close to the North Pole—

I was incredulous. I carefully read a scientific report out of Québec documenting high levels of PCBs and other chemicals in the breast milk of Nunavik mothers, chemicals they were passing to their babies in extraordinary concentrations. "How could this be?" I wondered. How could the Arctic, seemingly untouched by contemporary ills, so innocent, so primitive, so natural, be home to the most contaminated people on the planet? I had stumbled on what is perhaps the greatest environmental injustice on Earth.

I grew up, oblivious to such concerns, in one of the nation's toxic hot spots at the dawn of the environmental movement. I was a kindergartener when *Silent Spring* author Rachel Carson issued her dire warning about man-made chemicals—she called them "elixirs of death" —poisoning our world. Only much later did I learn that the Illinois suburb where I spent much of my childhood had the dubious distinction of being the PCB capital of the United States. Midway between Chicago and Milwaukee, on the shore of Lake Michigan, Waukegan had wide streets lined with glorious maple and elm trees and backyards big enough to play baseball in. But it also had a bustling harbor where one of the world's largest boat engine manufacturers was routinely dumping PCBs throughout my childhood, creating an immense, toxic mess. While I attended elementary school, the "Dirty Dozen"—the ubiqui-tous PCBs and chlorinated pesticides such as dichloro-diphe-nyl-trichloroethane (DDT)—were reaching record levels in all our urban environments, particularly around the Great Lakes. It was much later, in the 1990s, that I learned that children whose mothers ate a lot of fish caught in Lake Michigan had lower IQs and reduced memories. I often think back to summer days spent playing along the shoreline of the lake and family dinners at a popular fish restaurant overlooking Waukegan Harbor, and wonder, usually in the most abstract and impersonal of ways, what effects exposure to the contaminants of that era had on my generation. Use of these compounds peaked in the 1960s and contaminated all our foods, even our mothers' breast milk, so many people my age carry heavy body loads today. Perhaps these private musings are what has driven my conviction to understand and explain the risks of toxic chemicals, particularly to pregnant women and their newborns, the most sensitive of all populations. I was just shy of forty, and about to conceive my own child in another pollution hot spot, Southern California, when I learned that the bodies of Arctic women were carrying far more pesticides and industrial chemicals than my own, about ten times more than people like myself living in the shadow of skyscrapers and factories and polluted harbors. It turns out that the millions of tons of PCBs and other chemicals aren't staying put in places like Waukegan Harbor. They are hitchhiking long distances, poisoning remote, indigenous people in the most insidious way, through their traditional foods, their diet of whale and seal, which, for thousands of years has allowed them to survive in a place with no farms, no groves, no ranches.

I did not come to this project with an innate love of the Arctic. I was drawn there by the irony of its plight and my compulsion to explore and chronicle the damage inflicted on wildlife and people by contaminants spreading around the globe. Is any place on Earth safe? I knew the answer waited at the top of the world, where scientists are unearthing surprising revelations about how toxic substances threaten us all. The tale of the Arctic Paradox is an environmental whodunit, a scientific detective story, an anthropological journey, a chronicling of natural history, a lesson in biology and atmospheric chemistry of worldwide relevance, all wrapped together. It has all the elements of an engrossing novel—unexplained scientific phenomena, powerless and innocent victims, charismatic characters, beloved animals, international politics—all with an exotic backdrop—the harsh and haunting landscape of icy fjords, volcanic cliffs, and glacial tundra. The only thing this epic doesn't have is a villain. The villain, it turns out, is the whole rest of the world.

As I began my quest, I pondered what lessons the rest of us could glean from the Arctic's plight. I had a hunch that it would resonate with all Americans, all people everywhere, no matter where they live or what they eat. After all, even though the levels of PCBs, mercury, and other chemicals discovered in inhabitants of the higher latitudes are unparalleled, no one on Earth is untouched by these contaminants.

No life-form, virtually no food, is free of them. They spread throughout the web of life and build up in human body tissues over decades, mostly from consumption of tainted fish and other seafood. Once they enter our bodies, some of them never leave—except when they are transferred from mother to child through the womb and in breast milk. Although that notion is disheartening, I envisioned this as a story with a potentially happy ending. As the newborn child and polar bear in the Arctic are contaminated, so is the newborn child in Waukegan or Los Angeles. So if Arctic inhabitants manage to survive, so will we all.

On a June afternoon in 2001, in the extreme northern Thule region of Greenland, I took my first breath of air above the Arctic Circle. To my battle-scarred lungs, accustomed to the smog of Chicago and Southern California, it seemed the freshest air I had ever inhaled. I breathed deeply, letting its chilly dampness fill my lungs and its irony fill my thoughts. In Los Angeles, everyone knows it is unwise to breathe so deep but this, after all, was the Arctic, as pure as the driven snow. For the next two years, I chose to visit places—particularly Qaanaaq, Barrow, Svalbard, and the Faroe Islands—where I could see firsthand the impacts of contaminants on the indigenous people and animals; familiarize myself with this vast region's diversity of species and cultures; detail the effects of contaminants, both physical and cultural; and understand why people there are so reliant on a diet that leaves their bodies contaminated. Unraveling the mysteries behind the Arctic Paradox requires a voyage through time and space, spanning five decades and thousands of miles across the top of the world, from a dusty shoreline in Barrow, Alaska, where hundreds of Inupiat gather each spring to feast on giant bowhead whales, to an ice-sculpted fjord in Norway, where polar bears emerge from their dens with cubs no bigger than preschoolers but already loaded with chemicals. From the team of marine scientists baffled by missing sea otters along the foggy shoreline of Alaska's Aleutian Islands to the Inuit brothers guiding their sledges across Greenland's ice cap, an array of people have witnessed the altering of this wilderness. While journeying to the extreme north-ern end of the world, I also will lead you back in time, back to when synthetic chemicals emerged decades ago, and into the minds of some dedicated scientists as they try to decipher the subtle impacts on the health of Arctic people and wildlife, predict the region's future, and devise realistic solutions. My goal is to invoke a sense of place, a sense of time, a sense of character—and ultimately a sense of how all these things connect to give the Arctic realm a distinct personality that is undergoing a dramatic evolution as pollution from the rest of the world spills into the region. The Arctic is a place where the old blends with the new, with fractious results. The best way to portray this dilemma is through the eyes of its inhabitants—not through my own limited ex-periences. This is no travelogue, and although I will guide you, I aim to let the people and animals of the far North, and the scientists work-ing to protect them, tell their own tales.

I began this quest with more ignorance about the Arctic than af-fection, an uninformed fascination with a world so unlike my own. For centuries, the North Pole has captured the world's imagination and inspired a spirit of exploration and adventure, yet to most people it remains a mysterious, foreign place, indistinguishable from the Ant-arctic at the opposite end of the Earth. Most Americans are surprised to hear that anyone lives there except for a few Alaskan "Eskimos" in igloos. That was almost my starting point too, before I logged tens of thousands of miles by air, plus many more by dog sledge, ship, boat, and helicopter. Ancient world maps didn't even bother to recognize the Arctic. At the time of the fall of the Roman Empire, it was marked as uninhabitable, too cold to be civilized, and labeled "terra incognita," unknown land, as late as the 1600s. Yet a civilization has thrived there for five thousand years, perhaps much longer, harvesting the ocean and tundra of seals, whales, caribou, seabirds, and other prey. Although the imaginary Arctic Circle sits at latitude 66° 33' N, the circumpolar north is usually defined as encompassing sub-Arctic regions such as the Aleutian and Faroe Islands. Spanning three continents and divided among eight countries—Canada, Russia, the United States, Denmark/

Greenland, Norway, Finland, Sweden, and Iceland—the Arctic belongs to everyone, yet to no one. Largely forgotten and inevitably misunderstood, an estimated half a million indigenous people now inhabit the circumpolar north, and their population is steadily growing. Called Eskimos (translated as "eaters of raw meat" until the term fell out of favor as derogatory) the roughly 100,000 native people of Greenland, a territory of Denmark under its own home rule government, and of Nunavut and Nunavik, the Arctic regions of northeastern Canada, prefer to be called the Inuit or Kalaallit (both translated as "people"). Canada's Northwest Territories have the Dene and Métis, while its Yukon, next to Alaska, has the First Nations tribes. The Inupiat—some still call themselves Eskimos—live along Alaska's north shore, while the Aleuts and Yup'ik are farther south. The Chukchi, Nenets, and more than a dozen other tribes inhabit the Russian Arctic's vast Siberia and far east, while the Saami live in Scandinavia and on the Kola Peninsula in western Russia. The Faroese people, Danes in all aspects except for their diet of whale blubber, live on islands north of Scotland, just south of the Arctic Circle. Despite their varied heritage and scattered geography, all these people share the bond of being threatened by the far North's toxic contaminants. This pollution knows no borders, neither cultural nor political.

I visited five Arctic countries, coming within five hundred miles of the North Pole, reaching a northern latitude of around 80°. I shivered on naked floes and walked in hunters' footsteps to avoid shattering the ice and drifting off to sea. I ate whale countless times—raw, fermented, sautéed—and I came to crave boiled seal (best served with salsa) for the warmth and nutrition it provides. The Arctic preys on your weaknesses, and there were plenty of times when I was convinced that I wasn't cut out for this, that I didn't have the will or endurance to persevere through such rugged conditions. There was the time I spent camping on the ice with hunters in northern Greenland, five of us crammed into a makeshift tent, the fumes from an oil heater so overwhelming that I was sleepless the first night, wondering if I would die of carbon monoxide poisoning, until finally I convinced myself to trust the hunters who had spent many such a night like this. There were the days tracking polar bears in a helicopter when I got horribly airsick each and every time we swooped down, a Norwegian scientist hanging out the door aiming a tranquilizer gun at a bear, the blades whipping up a blizzard. I couldn't tell if we were crashing or landing so I just left my life in the able hands of the pilot and tried not to think of the other able pilots and scientists who had been forever lost in Arctic whiteouts. There was the time I tasted *mikigaq,* fermented whale blubber, an Alaskan delicacy so bloody, tough, and pungent that it's hard to believe it is considered food. Despite the risk of antagonizing my hosts, I had to spit it out. There was the time I learned, upon returning home to Southern California, that the ice broke beneath Inupiat hunting crews in Alaska the day after I had left them. There were the countless times my fingers were too frozen to jot down notes or push a camera shutter. These were the times—and they were frequent—that I dreamed of switching to a project investigating coral reefs in some calm, idyllic, tropical sea.

As merciless and forbidding as the Arctic wilderness is, I grew to crave its serenity and respect its raw power, and I never ceased to be amazed by the miraculous ways its people and animals survived. Initially, I thought Arctic life, particularly the diet of marine mammals, was an anachronism, a quaint but short-lived remnant of the past. But I quickly learned how wrong I was. Arctic inhabitants, throughout the centuries, have not only made do with what little nature gives them—they've excelled at perfecting it. They use traditional tools, eat traditional foods, and wear traditional clothes not because they are primitive but because they know that nothing else is as perfectly suited to their environment. I filled a suitcase with an expensive, bulky down parka, a survival suit, the Patagonia store's best fleece, and the newest innovation in boots that promised to keep my feet warm in temperatures sinking to 95 degrees below zero Fahrenheit. Wearing layer upon layer, I was bundled in a cocoon, barely able to move in all this gear, envious of the supple, simple sealskin boots and parkas worn by the Greenlandic hunters. I brought along PowerBars in case I couldn't stomach the hunters' food, but wound up throwing them out when they froze as solid as the ice beneath my feet. The trail mix that always served me well as a snack on hikes looked pathetic, and I welcomed the body-warming seal ribs boiling on a camp stove. Nothing in the modern world can match the survival skills and technologies of Arctic inhabitants. When Americans at an Air Force base in Greenland attempted to improve on Inuit sledges by substituting screws for strips of hide, their creations fell apart, too inflexible to bend with the ice. Gedion Kristiansen, one of the best hunters in Qaanaaq, tended to my boots to keep them dry and inquired often about my hands and toes to make sure they weren't frostbitten. "When I come visit you," Gedion said in his native Greenlandic, or *Kalaallisut,* translated by my interpreter, "then you'll take care of me." "It's a deal," I answered in English. But I knew that Gedion, like the rest of his family, had never left his homeland, and never would. One day, on our hunting trip, Kelly, my Inuit interpreter, told a joke. Two Inuit hunters from Greenland meet the king of Denmark. "This is the first time I've seen an Eskimo," the king says. "That's OK," one of the hunters says. "This is the first time I've seen a king." Gedion, his brother Mamarut, and Mamarut's wife, Tukummeq, bellowed with laughter. In my journeys, I often felt like that Danish king, an awkward alien, out of touch with the far North's native cultures. One day, Mamarut warned me to walk only in his footprints when we were out on the ice. Although he meant that literally, to make sure that I wouldn't fall through thin ice, I found it to be wise philosophical advice, too.

Greenlanders are a pragmatic people—everything serves a purpose—so they are befuddled by Americans' idealistic and imperialistic attitudes toward nature. Do Americans, asked Ujuunnguaq Heinrich, a Greenlandic minke whale hunter, actually spend money "adopting" wild animals? What purpose, he asked, is there in a slip of paper that pretends to let you own a whale or a seal or an otter? Why would anyone pay good money to adopt an animal that is free and wild and lives thousands of miles away in a place you will never see? Ujuunnguaq laughed at the notion, suggesting that perhaps he should adopt an American cow. Questions about my own culture and values surfaced often on my journeys. Why do Americans try to ban our hunts, when it's the only food we have? How can you say you worship whales when you are the ones contaminating them? I struggled to give answers that made sense but in most cases they sounded flimsy or self-serving. Lifestyles that seem so sophisticated to us seem so primitive to Arctic people. After all, even the most barbaric of animals know better than to foul their own nests.

On a spring day in Svalbard, Norway, when I knelt down to pet the fur of a four-month-old polar bear cub, its dark eyes were all innocence and wonder, the eyes of a baby. My son, Christopher, was four at the time, and I remember thinking that the cubs were his size and just as harmless, if not more so. Few visitors get the chance to see newborn polar bears in the wild because they den in the remote, inaccessible reaches of northern fjords, and at the time, it seemed as if I was visiting some frozen, surreal petting zoo. The two scientists and I were probably the only humans these bears would ever see in their lifetimes, yet I knew humans had already left an indelible, invisible mark on them. I feared for them, and their human counterparts—the countless generations of Arctic mothers and babies that are poisoned from afar, feeling the ill effects of pesticides and industrial chemicals that their society never produced, never used, never benefited from, never even heard of, simply because they live traditional lives. By and large, the Arctic's troubles are not self-inflicted. The pesticides we spray and the chemicals in our products are having profound effects on people and wildlife thousands of miles away. We are the polar bear's only predator, and our most powerful weapon isn't a hunting rifle—it's an old electrical transformer leaking PCBs.

Ingmar Egede, an Inuk who guided me in spirit on my journeys, taught me that not just their bodies, but also their souls are at stake when it came to contamination of traditional Arctic foods. On my first night in Greenland, in the contemporary capital of Nuuk, Ingmar stood in the kitchen of his apartment, slicing whale meat into bite-sized cubes. He scraped the meat, as blood red as cabernet, into a cast-iron pan and sautéed it in a bit of oil. He blended some paprika with milk for a sauce and added rice. Ingmar usually ate minke whale raw, straight from the freezer, the Greenlander's version of beef jerky. But this time he was serving dinner to an American guest. "It looks like beef stew," I said. Ingmar laughed. "Beef that has never eaten grass," he said. The son of a Greenlandic father and a Danish mother, born in Greenland's Disko Bay and educated in Copenhagen, Ingmar was much like Nuuk itself, a seamless blend of his homeland's ancient traditions and its modern conveniences. A grandfather with silver hair and silver-rimmed eyeglasses, looking a decade younger than his seventy years, he usually donned jeans and a black fleece jacket rather than the birdskin anorak he wore as a child. He had an exercise bike in his living room, an Apple computer in his den, and a narwhal tusk hanging on the wall over his dining room table. Shelves from floor to ceiling were lined with books in Danish and English about native peoples and animals around the world. He spent years in northern Greenland's most remote villages, teaching children, and even though he lived in modern Nuuk when I met him, his ties to the traditional way of life were strong. Ingmar served the whale with chopsticks and uncorked a bottle of Spanish wine. "Rioja goes well with it," he said. Over dinner, Ingmar and I discussed Jay Leno and American politics before veering off on the touchy subject of international attitudes toward whale hunting. Ingmar called Americans the most hypocritical people on Earth. No offense, he said. None taken, I said. The meat was quite tasty, and we each devoured a second helping. I ventured, awkwardly, to describe its flavor. Too intense to be beef. Nothing at all like fish. More like game. "Perhaps a bit like lamb?" I said. Ingmar frowned and shook his head. "Nothing is like whale. There is whale meat. And then there is everything else."

Upon finishing our meal, we shared some Belgian chocolates, and Ingmar lit a cigarette and decided to tell a whale tale. He was an eloquent storyteller, with a vivid vocabulary, bold gestures, and perfect, nuanced English. Once upon a time, Ingmar said, he was in a fishing boat, no bigger than a bathtub, in the waters off Alaska. A thirty-foot whale, perhaps a bowhead—he's not quite sure—appeared next to him. It raised its flipper and slapped the water, coming eye to eye with Ingmar, who introduced himself. "I'm Ingmar, a Greenlander," he told the whale in Greenlandic, the language, he said, that always brought him closer to nature. For an hour, the man and whale were companions. It may seem impossible for someone who eats whale to befriend one, but not so, Ingmar said. "I think a farmer would know the feeling—you respect the animals but it doesn't mean you shouldn't kill them. A whale is an extremely beautiful animal but one of the most beautiful aspects is it tastes so good and has a lot of meat," he said. He stressed that their diet offers far more than nutrition. "Our foods do more than nourish our bodies. They feed our souls," Ingmar said. When I eat Inuit foods, I know who I am. I feel the connection to our ocean and to our land, to our people, to our way of life. When I travel outside our homeland, my metabolism often goes wrong. Coming home and turning to Inuit food, I am all right again, within hours. When many other things in our lives are changing, our foods remain the same, and they make us feel the same as they have for generations."

Ingmar, an educator devoted to promoting human rights in the Arctic, knew it was difficult for outsiders to fully comprehend the strength of this ancient bond with words alone. So he took me to Nuuk's city hall. The building has all the trappings of twenty-first-century governance—computers, filing cabinets, multilined phones—and I thought he had escorted me there merely to show off Greenland's contemporary side. So I was ill-prepared for the splendor I saw when I walked into a conference room. Intricate wool tapestries designed by Nuuk artist Hans Lynge adorned its walls. Woven with the subtlety of watercolors, using natural fabrics and dyes, they beautifully portray everyday life in Greenland. One depicts a hunter returning to his village with a seal and the joy reflected in a child's face as his family shares the meat. In another, a kayaker caught in a storm makes his way safely home guided by an enchanted seagull, and in another, men proudly congratulate a boy after his first successful hunt. Their message is eternal; the same themes about the spiritual and cultural connection between man and nature have been reflected in Inuit art since the second millennium B.C. Looking through the artist's eyes, I felt as if I had peered inside the souls of Arctic people.

Later, Ingmar and I were aboard an old, peeling whaling boat in an icy fjord off the east coast of Greenland, observing a crew of three men hunting minke whales, when I realized that intellectually I can comprehend the Arctic ethic, but deep inside, in a soul crafted by urban comforts and disconnected from nature, I will always be a stranger there. I understand the Inuit's need to hunt. I defend their indigenous right to do it. I admire their culture and their will to survive. But all that empathy is shattered, for a moment, by the blast of a hunter's rifle. There is something very jarring about the sound of a gunshot at sea, a sound heard few places on Earth. In most of America, sailing out on the open ocean is one of the few places where you would never expect to hear that sound. For an instant, it wounded me. I cringed. It was purely a gut reaction, not a cognitive one, but I was puzzled and surprised by its force. I know the sea isn't a sanctuary; we use it and abuse it in countless ways. We trawl its bottom with nets, we pollute it with urban grime and chemicals, we wipe out species, we may be altering the climate, we spill tankers of oil. The taking of a whale or a seal for feeding Arctic people is perhaps the least harmful thing we do to the ocean as human beings. So why should I have such a powerful physical reaction to a single gunshot at sea? I contemplated this but I could not explain it. I glanced over at Ingmar. His aging face was luminescent as he enjoyed a whaling trip that, unbeknownst to all of us, would be his last before he died. He looked serene, as if he were worshipping at a church or temple. As a *Qallunaat*, a person of the south, I will never feel this sacred bond between man and prey, between man and his ancestors, yet I realized as I watched Ingmar that I must be able to decipher it and translate it if I am to tell this contemporary tale of the Arctic. As the echo of the gunshot faded across the fog-shrouded sea, my discomfort did, too. Icy rain fell like needles, stinging my face. I cinched up my anorak and waited.

13

PART I

THE ARCTIC PARADOX

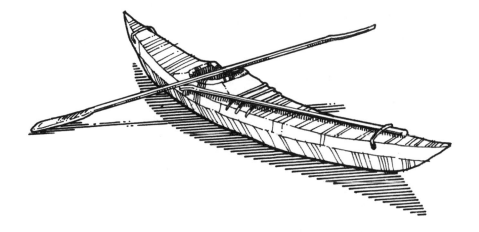

Chapter 1

Blowing in the Wind:
A Contaminant's Long
Journey North

When chemicals are spilled in urban centers, sprayed on farm fields, and synthesized in factories, they become hitchhikers embarking on a global voyage. Carried by winds, waves, and rivers, they move drop by drop, migrating from the cities of the United States, Europe, and Russia into the bodies of Arctic animals and people a world away.

The journey of a toxic hitchhiker might begin on a hot, steamy summer day in Chicago, along the shore of Lake Michigan. A fire sweeps through an office building, and the flames reach an electrical transformer, which catches fire and explodes. Thick clouds of smoke clog the sky, and oily liquids called polychlorinated biphenyls spray into the air and leak onto the pavement. These compounds, known as PCBs, were at one time among the most popular chemicals ever produced. First manufactured in 1929 by substituting chlorine atoms for hydrogen atoms in hydrocarbon formulas, PCBs do not burn and are nearly indestructible, so they were perfect coolants, insulators, and hydraulic fluids. During the first half of the twentieth century, electric companies purchased large quantities for immense electricity-storing devices like transformers. By the mid-1960s, many of the same characteristics that made PCBs ideal for industrial applications began to wreak havoc in the environment, particularly the Great Lakes, the Baltic Sea, and other industrialized areas. Their manufacture was banned in the United States and much of the world in the late 1970s. Yet thousands of old transformers still exist, about 21,000 of them in the United States alone that carry more than 100 million pounds of PCBs. Whenever they catch fire, leak, or are improperly disposed of, the compounds seep into the environment.

When this transformer explodes in Chicago, most of the PCBs, perhaps three-quarters of them, fall to the ground on city streets or drift into Lake Michigan. But some vaporize, turning into gases that float into the air. They spread out randomly in all directions. Some are lifted up toward towering columns of cumulus clouds, rising perhaps three miles up, where they are caught by fast-moving winds. Traveling at possibly 100 miles per hour, the PCBs quickly head east, crossing over Michigan and New York and then the Atlantic Ocean. The winds sweep them north, over Arctic Canada, then Greenland, then a remote string of Arctic islands in Norway. Within days, some of the PCBs from the Chicago transformer have moved across the top of the world, arriving near the North Pole.

As winter descends on Chicago, the PCBs left behind are trapped there, hibernating in a blanket of snow. But come spring, when temperatures warm again, they are set free, and they start globe-trotting. Earth's atmosphere is like a giant beer distillery, with continuous cycles of heating and cooling and condensing, and chemicals like PCBs are constantly seeking equilibrium in the environment, seeking out cold climates. They move from the air to the soil and the air to the ocean, and back again, in a phenomenon called the "grasshopper effect." When temperatures rise, the compounds evaporate in the heat, and drift along slowly in the atmosphere at a height of perhaps 1,000 feet. When temperatures cool, they condense—like drops of dew that form on grass—and fall to the ground. Some manage to move only a few hundred miles, from Chicago to Michigan, before they drop onto trees or roads or grass or cropland and remain there, slumbering throughout the cold winter. But come spring, when temperatures warm, they evaporate again, moving with northbound winds. Over the coming years, they continually rise and fall like this, hopping across the world, in search of a cold environment where they can eternally rest. Within a few years, they have joined the others that took a faster route to the North Pole. No one knows the precise amount of PCBs that are flowing to the Arctic, but by one estimate sixty-seven tons, mostly in the form of gases, arrive there every year.

Once there, the PCBs that reached northern latitudes via this atmospheric hopscotch—the fastest and most direct route to the Arctic—join others that took other, slower pathways. Oceans are powerful vectors, their currents slowly carrying masses of contaminants north. In John Donne's "For Whom the Bell Tolls," the poem that proclaims "no man is an island," a clod of dirt from Europe washes out to sea. "We now know where it goes," says Rob Macdonald of Canada's Institute of Ocean Sciences in Sidney, British Columbia. "It goes to the Arctic."

It might take years or decades for a drop of PCBs that originated in Chicago and fell into the Atlantic to flow to the Arctic Ocean, the smallest of the world's oceans, via the small passages between the continents. Some travel the wide, open waters between Greenland and Norway's Svalbard islands, but others squeeze in and out of the Arctic between narrow straits separating Canada from Greenland and Siberia from Alaska. The oceans, like the atmosphere, carry tons of contaminants northbound, largely from the United States and Europe. "The ocean is a big lumbering giant," Macdonald says, "but once it becomes the flywheel, it can become an important source of contaminants." Rivers also empty into the Arctic Ocean, unloading large volumes of chemicals, particularly from Russia. Drifting sea ice stores and transports them, too. There are even biological messengers—migratory birds, fish, and whales that move chemicals from place to place.

Arriving by all these various pathways, the globe-trotting PCBs permeate everything in the Arctic—its air, snow, ice, fog, soil, seawater, and ocean sediment—in all regions, no matter how remote, from Siberia to Greenland. Upon their arrival, most of the PCBs stop moving. About two-thirds of the PCBs that find their way to the Arctic stay there—perhaps forty-six of the sixty-seven tons are added to its environment every year, while twenty-one tons continue hopping around the world, according to one scientific estimate. These compounds are slow to break down in frigid temperatures, so they endure in the ice for decades, perhaps centuries. Arctic ice melts and freezes in endless cycles, and the chemicals tend to accumulate along the ice edge. This is where they end their physical voyage and begin a biological one.

If scientists were to measure only physical, or abiotic, elements, the Arctic environment would seem almost pristine, among the purest places on the planet. The Great Lakes, the Baltic, and the North Sea, for example, hold ten to one hundred times higher concentrations of PCBs than the Arctic Ocean. The air in Chicago contains a lot more PCBs than the air in Svalbard, Norway. So if the air they breathe and the water they drink and feed in are fairly clean, why do the bodies of the Arctic's people and top predators carry so much toxic trash? How contaminated an animal is depends not on where it lives but on what it eats—its place in the hierarchy of life, the food chain, or more accurately the food web. Capping a lush green and blue planet, the Earth's circumpolar north looks bleak and uninhabitable but in reality its 13 million square miles—as vast as the entire continents of North America and Europe combined—brim with an array of odd and hardy creatures, from ice-clinging, single-celled algae to polar bears, that can survive nowhere else on Earth. PCBs work their way through this ecosystem from the bottom up. They are absorbed first by sediment on the ocean floor and ice on the surface, and then they infiltrate single-celled plants that bloom during the spring, when light becomes available for photosynthesis. Each individual organism begins to build up the contaminants from its environment—a process called *bioaccumulation*. These plants, sometimes stretching into ten-foot-long filaments, are then grazed on by the ocean's zooplankton—tiny, shrimplike crustaceans such as copepods. From there, the Arctic's food chain spreads into a vast, intricate web, reaching out in multiple directions. The copepods are eaten by cod, the cod are eaten by toothed whales such as narwhals, and the narwhals are eaten by Inuit hunters. Ringed seals eat the cod and then polar bears and human hunters eat the seals. Walrus feast on the copepods, people feast on the walrus. Seabirds eat cod, people eat the birds.

In the Arctic, you are what you eat. Because PCBs are not easily expelled from an animal's body, they accumulate there—this means that animals at the top of the food web have eaten all the contaminants consumed by their prey and their prey's prey. Arctic food ladders have as many as five rungs, and at each step up, the chemicals can magnify in concentration twentyfold or more in a phenomenon called *biomagnification*. Occupying the top rung are people and polar bears, which can carry millions, perhaps billions, of times more PCBs than the waters where they harvest their foods.

This trip through the food web, in most cases, occurs via the fat of animals. PCBs are not water-soluble, they are lipid-soluble, which means they are readily absorbed by tissues with a high fat content, and they cling there, rather than dissolving and flushing out of the body in liquids such as urine. "If you were to ask a PCB molecule where do you like to be, it would say it likes to be in the company of molecules similar to itself. As it looks around, the molecules that are closest in chemical nature to PCBs are fats, or lipids. So PCBs tend to accumulate in fat," says Donald Mackay of Trent University's Canadian Environmental Modelling Centre. In cold climates like the Arctic, sea mammals have an unusual propensity to produce fat. Their blubber is an insulating layer several inches thick, and their milk is mostly fat. Fat stores energy, but it also stores a variety of man-made chemicals, which makes Arctic mammals more susceptible to the buildup of contaminants than leaner animals in temperate zones. Making matters worse, many Arctic creatures experience seasonal cycles of fattening and starvation, and when they use up their fat reserves in winter, PCBs concentrate and migrate into their vital organs. Large Arctic predators such as polar bears and whales also tend to live longer than land animals so their bodies store many decades' worth of chemicals, increasing in concentration with their age.

A few decades ago, the solution to pollution was dilution—scientists thought that compounds dumped into the vast waters of the oceans would be out of the way and rendered harmless. What they didn't understand was the underlying biology of the sea—that chemicals have a far greater potential to accumulate in ocean life than in land-dwelling creatures. Author and biologist Rachel Carson, in her 1962 classic *Silent Spring,* warned of the "sinister" nature of chemicals that are passed from one organism to another. She knew that DDT sprayed on alfalfa wound up in the livestock that ate it, and ultimately in the people who ate the livestock. But ecosystems on land are simple—a cow eats grass, people eat the cow. In contrast, the Arctic Ocean has a long and complex food web, with so many layers that toxic compounds build up to extraordinarily high concentrations.

Every human being, no matter where on Earth, contains traces of these toxic compounds because of the chemicals' ability to persist in the environment and magnify in the foods everyone eats, particularly seafood. But Arctic people are especially vulnerable because of their place at the very top of the natural world's dietary hierarchy. They eat 194 different species of wild animals, most of them inhabiting the sea. Often on a daily basis, they consume the meat or blubber (*muktuk, mattak,* or *maktak* in Arctic languages) of fish-eating whales, seals, and walrus four or five links up marine food chains. In contrast, many urban dwellers have lifestyles that distance them from their polluted environment. Most have abandoned a hunting culture for agriculture, eating much lower on the food web with a diet of mostly land-raised vegetables, grains, beef, and poultry that contain less contaminants.

PCBs aren't the only hitchhikers determined to settle in the Arctic. Like PCBs, the chlorinated pesticide DDT was first synthesized in the 1800s, although it wasn't used as an insecticide until the 1940s, after Swiss scientist Paul Hermann Müller noticed its ability to kill pests. Müller's discovery was considered so miraculous that he later won a Nobel Prize for medicine. But DDT began building up in the environment worldwide, wiping out birds and other creatures. By the early 1970s, its use was banned in North America and Europe. Still used in some countries to fight malaria and stockpiled in others, it is still spreading globally, reaching the Arctic by the same atmospheric and oceanic pathways as PCBs.

Today, about two hundred toxic pesticides and industrial compounds have been detected in the bodies of the Arctic's indigenous people and animals, including all twelve of the "Dirty Dozen," the so-called "legacy" organic pollutants such as PCBs, DDT, mirex, dieldrin, and chlordane that are capable of inflicting the most ecological damage. They are joined by mercury, a potent neurotoxin released by coal-burning power plants and chemical factories. Mercury is on the rise in many animals of the Arctic, and so are a variety of new contaminants such as brominated flame retardants, widely applied to plastics and foam, and perfluorinated acids, formerly used in Scotchgard and still used in making Teflon. Unlike PCBs, DDT and brominated flame retardants, which accumulate in fat, mercury and perfluorinated chemicals build up in protein-based tissues such as the liver.

Because these synthetic marvels can survive virtually anything they encounter on their global voyages, they are gradually building up in the remote reaches of the far North to levels that jeopardize people, wildlife, and cultures that have survived this harsh environment for millennia. The Arctic's people and animals have been transformed into living, deep-freeze archives storing toxic memories of the industrial world's past and present. This phenomenon is so insidious that no one tapped into it for decades, until one day thirty-two years ago when scientists stumbled on silent messengers that came in the form of plump Canadian seals.

Chapter 2

Unexpected Poisons:
Serendipity at the Top
of the World

Back in the early 1970s, Tom Smith knew more about seals than any white man in Holman, a treeless expanse of frozen tundra that is one of the northernmost outposts in all of Canada. He knew when the seals had pups, how long they lactated, how big they grew, what they ate. Most of what he learned as a field scientist he gathered with the help of Inuit hunters, who camped on the sea ice of Canada's Northwest Territories every winter, enduring twenty-below Fahrenheit temperatures as they waited for ringed seals to pop through their breathing holes. The few hundred people of Holman, who hunted seals for food, skins, and oil, still lived like their ancestors did and had little contact with the rest of the world. Sometimes Smith would persuade the hunters to share their prey with him, and he used the blubber to explore whether Arctic seals were healthy and well-nourished.

One day Smith read a piece in an obscure scientific journal written by a chemist in Nova Scotia who had tested harp seals in the waters off Québec's urbanized coast. The chemist, Richard Addison, had discovered toxic pesticides and industrial compounds in the urban seals' blubber. It was 1973, an era of extraordinary pollution, when DDT, PCBs, and other compounds were building up in animals throughout North America and Europe. Many birds—eagles, robins, pelicans—were vanishing as the chemicals destroyed their eggs. More curious than concerned about his seals, Smith thought that the specimens he had been collecting in Holman could be a treasure trove for a chemist since no one had ever tested the seals before. So he called Addison and asked: Would you like some Arctic blubber?

Addison had never been to Holman. In fact, he had never been to the Arctic at all. A few years earlier, in 1969, a Scottish fisheries scientist, Alan Holden, had found residue of DDT and PCBs in a few Arctic seals from Norway's northern coast and Canada's Baffin Island. The amounts were minute, almost undetectable, and Holden had declared them "substantially free of contamination in all but the 'background' sense." With the Holman seals inhabiting such remote waters of the Beaufort Sea, Addison thought it was likely that they, too, would essentially be "blanks," carrying no toxic substances or mere traces. Nevertheless, on a whim, Addison told Smith: Sure, send them and I'll take a look.

Soon afterward, blubber from about forty Arctic seals arrived at Addison's lab in Dartmouth, Nova Scotia, wrapped in foil and still frozen. It offered the first tangible clue to toxic detectives that chemicals were invading the far North.

Addison knew a lot about animal fat. Growing up Belfast, Northern Ireland, he had gotten his doctorate in agriculture, specifically studying the fat intake of poultry. He was most interested in applying chemistry to the real world, and when he was offered a job in Halifax, Canada, working in the "lipids group" for the government's fisheries research board, he accepted. He knew nothing about fish but he thought it would be a chance to help Canada figure out some commercial uses for blubber and fish oil. He started work there in 1966—when the environmental age was in its infancy. Chemical crises were just beginning to unfold around the world. PCBs were being detected in Swedish fish and Great Lakes birds. DDT was building up in California seabirds. As a chemist working with fish, Addison soon found himself completely drawn into Canada's emerging environmental problems.

In January 1969, a large detergent factory had opened along the shore of Newfoundland, and within two months, local fishermen noticed thousands of dead herring. Divers were sent down and reported that everything in the harbor was dead. It was an ecological and economic disaster—the first pollution event in Canada attracting national attention. The government directed Addison to investigate. It was a time when environmental chemistry didn't even exist, when environmental science of all types was in its infancy and there were few laws governing industry's handling of chemicals. Addison saw huge amounts of effluent flowing from the plant, a brew of unknown chemicals. Developing new detection techniques for seawater, he identified the culprit in the fish kill—phosphorus from detergents. The plant was shut down and the harbor floor dredged and paved over.

After that crisis, Canada decided to open up a pollution lab in Dartmouth, and Addison, hooked by the new science, took a job there in 1971, with the goal of developing ways to measure the new "bad boys" of pollution, organochlorines—chlorinated chemicals such as DDT and PCBs that had just begun showing up in wildlife. Addison knew he needed to push the limits of old-fashioned detectors in searching the environment for these chlorinated compounds, but he wasn't sure which medium to measure. Most scientists had been sampling water. He could have chosen fish or plankton but he decided on something he knew—fat. He needed large reservoirs of it for the sampling. What could be better than a plump seal with its thick layer of blubber? Canada certainly had lots of seals, and they were fairly easy to catch. He set out to turn seals into a sentinel species, an environmental monitor for the health of the whole oceanic ecosystem. He started in the foul waters of the Gulf of St. Lawrence off Québec.

Science is often serendipity, and in this case, Addison recalls, "the Arctic connection was purely accidental." He never would have thought to sample Arctic animals if Tom Smith hadn't seen his report and happened to pick up the phone. Addison was intrigued but not alarmed by what he detected in the blubber Smith sent: The levels of DDT and PCBs seemed pretty benign, an order of magnitude lower than animals in urban environments such as the Great Lakes. In 1974, Addison published a concise, three-page report documenting that the males were more than twice as contaminated as the females, a sign that the mothers were offloading the chemicals to their pups in their milk. A year later, two Canadian scientists reported PCBs in another Arctic species, polar bears.

At their labs in Scotland and Canada, Holden and Addison suspected that the chemicals they found in the Arctic were coming from distant, urban lands. How else could pesticides and PCBs be turning up in northern latitudes, and even in Antarctica? And what else could explain why DDT was decreasing rapidly in urban seals but not in Arctic ones? As early as 1969, Holden sounded a warning, the first of many to come: "One further aspect . . . should be emphasized," he wrote at the conclusion of a paper written for a meeting of European marine scientists. "The global distribution of the organochlorine contaminants in the marine environment has been demonstrated. These contaminants, particularly the PCBs, are chemically very stable and presumably other substances of similar stability will also be globally distributed." This contamination, Holden wrote, "could be potentially damaging" to Arctic seals and other animals, just as it has been to urban birds.

It was a prophetic warning—that large volumes of chemicals were spreading globally—but no one in the scientific community or the public heeded it at the time. The early discoveries about one of the most isolated places on Earth were promptly forgotten. No one bothered to follow up on them for nearly a decade. More pressing environmental crises were mounting in cities around the world in the 1970s. In the United States, oil gushed from rigs, rivers caught fire, skies were blackened by soot and smog, songbirds dropped dead, and many species were on the verge of extinction. The Great Lakes and Europe's Baltic and North seas were far more contaminated than the Arctic Ocean. Scientists were busy testing for chemicals in cow's milk and beef and chicken and butter and eggs and fish. Why should anyone care about a few seals or bears near the North Pole? It didn't dawn on them that there was another creature precariously perched at the very top of the food chain, eating marine mammals and passing the chemicals to its young. No one gave a second thought to the Arctic's human hunters.

When the first reports about seals and polar bears were published, Derek Muir had just begun work on his doctorate in agricultural chemistry at McGill University in Montreal. By the late 1970s, when Muir worked at Canada's fisheries department, organochlorines were no longer the new, hot curiosity attracting environmental scientists. Muir, like most of his colleagues, thought, "Why bother with chemicals like PCBs and DDT and chlordane?" The United States, Canada, and Europe—the most contaminated places—had just banned their production, so there seemed little reason to keep studying them. But, in the early 1980s, along came an intriguing advance in technology—a sophisticated type of gas chromatograph that was extremely sensitive in detection of the 209 varieties of PCBs and other organochlorines. It allowed chemists for the first time to figure out which specific compounds were in animals, giving them insights into their diets and metabolism and the sources of the contaminants. Muir couldn't resist. He had always been interested in the fate of chemicals. Even as a child he wondered what happened when he flushed an aspirin down a toilet. Muir teamed with his college classmate, Ross Norstrom, at the Canadian Wildlife Service in Ottawa, and they decided to follow up on the polar bear discovery, which had been neglected for almost a decade. Norstrom, who began the sampling in 1981, remembers that no one seemed to care about Arctic animals or worry about how PCBs or pesticides like DDT, hexachlorocyclohexane (HCH), chlordane, or toxaphene got there. His colleagues, focused on the Great Lakes, acted as though he were investigating alien beings on some distant planet. Using samples collected from a variety of locations near remote Baffin Bay, Norstrom and Muir were the first to trace toxic substances through the Arctic food web—from cod to seal to polar bear. When their detailed data were published in a journal article in 1988, their colleagues were stunned to learn how quickly the chemicals built up to high levels in marine life. Between fish and seals, some pesticides increased as much as sixty-two-fold, and forty-seven-fold between seals and polar bears. They even found one—chlordane, used on corn, citrus, and home lawns and gardens—in Arctic animals but not in marine animals farther south, suggesting that there was more chlordane in the Arctic than in the cities and farms where it was actually used. It defied explanation, as no one in the Arctic ever sprayed insecticides, the animals never migrate, and much of North America's chlordane production had ended a decade earlier. It was the first realization that the so-called background levels in the Arctic were not mere residue. A variety of chemicals had reached high levels in top predators and the phenomenon was as yet unexplained. The work by Muir and Norstrom gave scientists a baseline—a starting point—to determine effects on seals and bears. "No one had ever looked at the fourth-level predator before," Muir says.

At virtually the same time, in a lab in Québec, a physician had stumbled on the fifth-level predator.

Eric Dewailly was a teenager about to graduate from high school in northern France when he visited Africa's poverty-stricken Ivory Coast. His father was a doctor, a gastroenterologist, but Dewailly had no interest whatsoever in medicine. Instead, he was fascinated by the forces that made life in isolated Third World places so difficult, and he expected to become a sociologist. By chance, while there, he met a physician handling infectious diseases and visited a clinic, watching the doctor take preventive steps that actually saved people's lives—administering vaccines, cleaning up sewage, finding clean drinking water. He was so impressed that he decided to enroll in medical school, and he knew immediately that protecting public health was his calling.

Dewailly thought tropical medicine would be his specialty, but instead, because of an exchange program between France and the French-speaking Canadian province of Québec, he wound up in the opposite hemisphere. When he was asked in 1983 to return to Québec

City to start an environmental health program at Laval University, he jumped at the chance. At the time, chemicals were being discovered in the breast milk of women in the United States and Europe, but little was known about Canada. The Québec provincial government asked Dewailly to survey women and, in 1986, he chose women giving birth at twenty-two hospitals, mostly from around Montreal, which he assumed would have the worst contaminant levels.

As his project began, he happened to meet Johanne Gagnon, a midwife from East Hudson Bay, at a public health meeting in June of 1986. Gagnon asked him if he would like to include women in Nunavik, the Arctic region of Québec, home to about eight thousand Inuit. At first, he had little interest. Too many logistical nightmares. And the milk of women so far from industries would most certainly be pristine. Nevertheless, he agreed, thinking a few samples might be useful as blanks so he could compare an unexposed population to an urban one.

About a year later, in the fall of 1987, the first batch of samples from Nunavik—glass vials holding a half-cup of frozen milk from each of twenty-four women—arrived via air mail at the laboratory in Québec City. Lab technician Evelyne Pelletier removed a sample from a walk-in refrigerator and began the daylong preparations to analyze it. She first extracted the chemicals by adding a solvent compound to the breast milk and shaking it, then mixing in an acid to destroy the fat so it wouldn't plug up the instruments. She poured the organochlorine mixture into a narrow glass column, removed the impurities, and spun it in a centrifuge so that the liquids evaporated and only the highly concentrated chemicals were left. The next day, using a gas chromatograph, Pelletier screened the extract for twenty-two chemicals—ten insecticides and twelve PCB compounds. She stood at the chart recorder as the machine spit out reams of data, one chemical at a time. Within minutes, Pelletier knew something was wrong. The concentrations of chemicals were off the charts—literally. In a normal test, technicians find individual, needlelike peaks, like those on an electrocardiogram. Instead, the peaks had overloaded the lab's equipment, running off the page. Pelletier showed the charts to the lab director,

Jean-Philippe Weber. He had never seen his lab's equipment over-loaded by a sample. The concentrations were about thirty times higher than anything he had ever seen before. When he saw that the samples were the milk of Arctic women, making the results even more improbable, he called Dewailly. "We have a problem here," he said. Something was wrong with the Arctic milk. He thought it might have been tainted in transit with some type of solvent.

They decided to test the sample again, diluting it this time to get a more accurate reading, and then tried another batch of Arctic milk, and another. "We knew then that this was not accidental contamination," Weber said. The chemicals were real. They were the same contaminants found in the milk of women in the south—PCBs and pesticides—but the milk of the Arctic mothers had up to ten times more than that of the mothers in Canada's biggest cities.

To Dewailly, who grew up near the North Sea, in one of Europe's most polluted regions, it belied all logic—until he began to search ecological journals and unearth data about PCBs and DDT, including Addison's long-forgotten 1974 report about seals. It became clear that the seals of Holman had been an unrecognized omen. Dewailly knew that the Inuit ate marine mammals but, like most doctors, he had no idea that toxic substances were building up in Arctic animals, as the data were not published in the medical journals he read, only in ecological journals he had never even heard of. Dewailly contacted the World Health Organization in Geneva, where an expert in chemical safety told him that the PCB levels were the highest he had ever seen. Those women, the expert said, should stop breast-feeding their babies—immediately.

Dewailly hung up the phone, his mind reeling. He knew that no food is more nutritious than mother's milk and that Nunavik is so remote that mothers had nothing else to feed their infants. As a doctor, he couldn't, in good conscience, tell them to stop breast-feeding. But he couldn't hide the problem either. "Breast milk is supposed to be a gift," Dewailly says. "It isn't supposed to be a poison." At the same time, elsewhere in Canada, in a small Inuit community on Baffin Island, other medical researchers were finding similar levels of contamination in breast milk there. The news about Nunavik and Baffin Island spread to the highest levels of government in Canada in 1988, frightening and angering Inuit leaders and triggering an international investigation into the health of all Arctic inhabitants.

Dewailly, teaming with a doctor in Greenland, soon discovered that the bodies of some Inuit there carried such extraordinary loads of chemicals that their bodies and breast milk could be classified as hazardous waste. Over the next decade and a half, Dewailly led a team investigating the effects on the babies of Nunavik. He discovered that the Inuit's traditional diet of seal meat, beluga blubber, and walrus is part tonic, part poison: Rich in nutritious fatty acids, the foods protect the Inuit from cancer and heart disease but the research suggests that they also make babies more susceptible to infectious diseases and damage their developing brains. Nevertheless, Dewailly still firmly believes that the Inuit should keep nursing their babies and eating their traditional foods. Even today, almost two decades later, Canada remains embroiled in a debate over how to protect the health of its aboriginal people from the extreme levels of contaminants.

At about the time Dewailly began testing breast milk in Nunavik, Pál Weihe, across the Atlantic, was wondering about the childen of his own homeland. The son of a harbormaster, Weihe was born in the tiny seafaring village of Sørvágur in the Faroe Islands, a Danish territory in the middle of the North Atlantic, south of the Arctic Circle. In 1969, when he graduated high school, he left for Copenhagen to become a doctor—a surgeon, he thought. But upon studying occupational medicine, he learned about the dangers of chemicals and found this field more intriguing than his surgical studies. Surgery carries few surprises, he decided, but determining the risks of chemical exposure was so full of uncertainty, so mysterious, yet so vital to public health. In 1985, soon after his second child was born, Weihe heard about high levels of mercury in whales of the North Atlantic. The Faroese people are Nordic, not Inuit, but one thing separates them from the rest of their Danish compatriots: They eat pilot whales, in a tradition dating back centuries, perhaps to the days of the Vikings. Sometimes called "black torpedoes," pilot whales migrate long distances along the shores of the Atlantic, accumulating excessive levels of mercury from a variety of sources, including emissions spewed by coal-burning power plants thousands of miles from the Faroe Islands. The evidence that mercury is a neurotoxin that scrambles the brain dates back at least two centuries. "Mad as a hatter," a phrase made famous in *Alice in Wonderland,* originated from the tremors, confused speech, and hallucinations of nineteenth-century hatters poisoned by mercury used to cure felt. Yet it wasn't until the 1950s when the dangers to an infant's developing brain became apparent. At Japan's Minamata Bay, where a chemical factory dumped tons of mercury, thousands of people died or suffered various degrees of brain damage from eating fish, and an unexpected impact surfaced with the next generation: Thousands of children were born with mental retardation, deformed limbs, and other severe problems. Iraqi children suffered a similar fate in the 1970s when grain was contaminated with mercury.

Weihe knew Faroese babies were not exposed to enough mercury to cause retardation as in Minamata or Iraq. Nevertheless, he wondered if there were more subtle neurological effects at the lower exposures of his fellow islanders. Weihe approached Philippe Grandjean, a Danish environmental epidemiologist known at the time for his studies of another heavy metal, lead, which damages the brains of babies and children. Weihe told him about the pilot whales. Grandjean's response reinforced his concerns. "I'm afraid mercury could very well be like lead," Grandjean said.

Weihe returned to the Faroes as medical director at its hospital system and began to collect blood samples from adults. Sure enough, their bodies contained large amounts of mercury. Weihe and Grandjean mapped out an ambitious plan: Even though they knew it would mean a lifetime of work, they decided to assemble a group of subjects, called a cohort, and follow them from birth. Their first grant proposal was denied but the Danish government gave them $15,000 in support, enough to get started. They asked pregnant women throughout the Faroe Islands to participate. Very few said no. They wound up recruiting 1,023 pregnant women—80 percent of the women who gave birth on the islands in 1986 and 1987. Their umbilical cord blood was stored for mercury analysis.

When the babies were born, Weihe and Grandjean saw no indication of any immediate damage to the infants. But they didn't expect to. Maturation of the brain is what's at stake with lower levels of mercury exposure. Weihe and Grandjean decided to test the children when they reached seven years, the age school begins in the Faroe Islands. In the spring of 1993, the children underwent an extensive series of psychological and neurological tests designed to see whether the mercury impaired any of their mental skills. The results: a measurable delay in transmission of signals from the ears of the most highly exposed children to their brains, a subtle slowing of a key neurological function. It was the "eureka" moment, Grandjean recalls. Other tests on the children found impaired vocabulary, memory, and attention span at what had previously been considered a low and safe level of mercury in pregnant women. Grandjean and Weihe wrote up their first findings in 1994 but the scientific paper came back three times from the publisher for more review because the findings were so worrisome for seafood eaters around the world. The first results were finally published in 1997. Years later, in 2004, results of tests on the children when they reached age fourteen were published, suggesting that at least some of the neurological impacts of mercury are long-lasting, perhaps permanent.

After having tested nearly 2,000 Faroese children, Weihe now is convinced that the effects are real. As a scientist, he finds the research, which has had international repercussions for setting health standards, exhilarating, but as a doctor with strong ties to the people of his homeland, he is dismayed. Unlike the Inuit of Canada and Greenland, women in the Faroe Islands are now advised, based on Weihe's recommendations as head doctor of the hospital system, to stop eating the whale meat and blubber that have been important to their culture for centuries. It doesn't make Weihe, now fifty-five years old, the most popular person in the islands. But almost 2,000 mothers have come to trust Weihe, the "mercury doctor," with the well-being of their children, returning year after year to his small clinic in a residential neighborhood of Tórsham to undergo the neurological tests.

"My first assumption, back in 1985, was that we would not find any effects, that we have adapted to our diet over hundreds of years," Weihe says. Now he knows he was wrong.

Jim Estes was trying to solve a mystery of his own on another string of sub-Arctic islands half a world away, between the North Pacific and the Bering Sea. A marine biologist at the U.S. Geological Survey (USGS) in Santa Cruz, California, Estes for years had devoted his career to studying sea otters inhabiting the ocean off central California, trying to figure out why they had such a high death rate. Could pollutants like DDT and PCBs be to blame? After all, everyone knew California's waters were contaminated. For the sake of comparison, in the summers of 1991 and 1992, he traveled to the place where everyone assumed sea otters were clean, healthy, and thriving: Alaska's Aleutian Islands. He returned to Santa Cruz with ice chests containing samples of blood and fat from otters on Adak Island, and asked Walter Jarman, a pollutants expert then at the University of Utah, to take a look at their chemical content. The day the lab results came back, Jarman scanned the columns of numbers. No way, he thought. The Aleutian otters were supposed to be the uncontaminated ones, but he had never seen PCB numbers so high. How could otters inhabiting these remote Alaskan islands contain twice as much of these industrial compounds as otters off urban California? They carried 309 parts per billion of total PCBs—thirty-eight times more than otters in southeast Alaska. At the same time, another USGS scientist, Robert Anthony of Oregon, was finding surprisingly high concentrations of DDT and mercury in the eggs of bald eagles nesting in western parts of the

Aleutians. Most baffling of all, neither the otters nor the eagles ever migrated. They were somehow picking up the chemicals without ever leaving the islands.

Within a few years, by 1997, Estes had an even bigger mystery on his hands. The ecosystem of the Aleutians had collapsed. Tens of thousands of the Aleutians' otters disappeared within a matter of years, bringing them perilously close to extinction. There were no bodies to dissect, no clues to decipher. The otters weren't starving. They weren't sick. They simply vanished from the Aleutians without a trace, along with other sea mammals, particularly seals and sea lions, that had begun their descent in the 1980s. Estes believes that contaminants are not the main culprit, although they may play some minor role. Instead, he and his team have collected clues suggesting that various other human impacts, dating all the way back to commercial whaling a half-century ago, have triggered a series of events that upset the region's ecological equilibrium, upending its balance of predator and prey.

Piece by piece, Estes and other scientists think they are cobbling together this intricate puzzle, although the answers are more disturbing than satisfying, more elusive than conclusive. Estes predicts that the Aleutians' otters, in all likelihood, will be extinct in ten years, and their loss will reverberate throughout this entire ecosystem. It seems the ocean's chain of life is actually a fragile silk web. If you remove a strand, the whole thing unravels. And it may never be whole again.

In the late 1980s, scientists worried that the effects of contaminants in the Arctic were too subtle to see in its wildlife but that, slowly, over time, its populations could be decimated. They realized that they needed to probe the bodies of animals for minute biochemical changes. By the 1990s, wildlife researchers, using new techniques honed in part by AIDS researchers, had succeeded: They developed technology to look at specific parameters—such as immune cells, antibodies, or testosterone levels—and see if they changed with concentrations of toxic chemicals in the bodies of animals. By searching for such connections, they could finally answer the question: Were chemicals harming Arctic wildlife in ways impossible to detect with the naked eye?

The answer came on Svalbard, a Norwegian archipelago where Canadian Andrew Derocher was pursuing his dream of studying animals in remote, pristine lands. Coming of age in the heyday of the environmental movement in the early 1970s, Derocher was the only one in his family interested in the outdoors. He was raised along the lush banks of British Columbia's Fraser River, where he collected bird eggs and garter snakes and fished for salmon fry, trying to keep them alive in jars. After high school, he took a job as a seasonal game warden in British Columbia's provincial parks, spending his days fly-fishing for trout as bears ambled to the banks of the river. Then he studied wildlife ecology, focusing on large mammals, particularly bears. His father, a telephone repairman who worked his way up to upper management of a telephone company, was skeptical of his son's chosen career. His other son was a doctor and his daughter was a nurse, while his middle child was "mucking around with bears." When he took a research job at the Norwegian Polar Institute in 1996, Derocher thought he had found polar paradise. His longtime dream as a biologist was to study polar bears in their purest form, to find a population protected from human contact. Hunting of Svalbard's bears dates back to the sixteenth century—several hundred were trapped and shot for their fur each year—but since 1973, the archipelago has been a revered national refuge where hunting is banned. When Derocher arrived a quarter-century later, the population should have been fully recovered.

It wasn't long before he knew something was wrong. "Things just don't appear right," Derocher told his colleagues. It was as if these bears were still being hunted. Why weren't there more bears? Where were the older ones? Why were there so few females over the age of fifteen bearing cubs? Were they dying? Were they infertile? "Within the first year, it became pretty darned clear that I wasn't working with an unperturbed population," he says. In his second season on Svalbard, Derocher checked the sex of one bear as he routinely did, and found both a vagina and a penis-like knob. "What the hell is this?" he thought.

He had examined more polar bears than just about anyone on Earth, yet he had never seen that before. Then he started finding more—three or four out of every one hundred examined. Derocher immediately suspected that chemicals were to blame.

By then, it was indisputable that Svalbard's bears had extraordinarily high concentrations of PCBs. In living animals, worse doses had been found only a handful of times: in Pacific Northwest orcas, European seals, and St. Lawrence River belugas. A few years later, in 2004, Derocher and Norwegian scientists published some groundbreaking findings, documenting an array of effects in Svalbard's polar bears they linked to the PCBs. Included are altered sex hormones—reductions in testosterone, increases in progesterone—as well as depleted thyroid hormones, which regulate brain development of a fetus. The bears also suffer suppressed immune cells and antibodies; altered cortisol, which is important for managing stress and crucial body functions; lower retinol (vitamin A), which controls growth; and even osteoporosis. Such biochemical changes could impair the bears' ability to fight off disease and give birth to healthy cubs. Polar bear scientists theorize that the chemicals are culling older bears and weakening or killing cubs, perhaps leaving a missing generation of mother bears. Only 11 percent of Svalbard's bears with cubs are over fifteen years old, compared with 48 percent in Canada. When it comes to the most dramatic discovery—the pseudohermaphroditic polar bears with female and partial male genitalia—some scientists now suspect that they are natural occurrences, unrelated to the contaminants. Derocher and others, however, say that contaminants are a more plausible explanation. Essentially, no one really knows.

Nevertheless, Derocher is now virtually certain that there is some connection between the low numbers of bears in Svalbard and the toxic chemicals. "Could you realistically put two hundred to five hundred foreign compounds into an organism and expect them to have absolutely no effect?" he says. "I would be happier if I could find no evidence of pollution affecting polar bears, but so far, the data suggest otherwise."

Today, more than thirty years after the first traces of DDT were found in the Canadian seals, the evidence is overwhelming that toxic substances have spread throughout the Arctic, harming animals and people of the far North. An international body of scientists called the Arctic Monitoring and Assessment Programme (AMAP) concluded in a 2002 report that the contamination raises "fundamental questions of cultural survival, for it threatens to drive a wedge of fear between people and the land that sustains them."

Several generations have passed since chemicals first hitched a ride to the Arctic around World War II. The hunters who ate the seals sampled by Tom Smith are now likely to be grandparents, and the infants who drank the breast milk sent to Eric Dewailly's laboratory in Québec are teenagers, about to bestow on their children the chemical load amassing in their bodies from their consumption of marine life. Yet little has changed to protect the next generation of Arctic children.

After a half-century of research, scientists now ponder why it took them so long to make the connection between the Inuit and their prey, and to realize that toxic chemicals can wreak subtle damage on animals and people. Arriving in the 1940s and discovered in the early 1970s, Arctic contamination was ignored until the late 1980s, and by then it was too late. It had reached extraordinary levels throughout the circumpolar north. "We knew that there were contaminants of concern in the Arctic as early as the 1970s. Why it took twenty years for the other shoe to drop is a puzzle," says Rob Macdonald of Canada's Institute of Ocean Sciences. "There are many we could fault—maybe scientists, maybe politicians, maybe society. We could point fingers in lots of directions. But I think it's probably more instructive to say that sometimes we don't pay attention to things that we should, that we don't connect things well."

Making matters worse, contaminants aren't the only environmental threat to the Arctic. It faces a triple whammy of human influences—contaminants, climate change, and commercial development—that the

United Nations Environment Programme says is likely to inflict drastic changes on its natural resources and way of life this century. Seemingly hearty, the Arctic is, in fact, fragile.

What does this portend for the health of the world's children and animals, particularly in the Arctic, with its extraordinary exposure and vulnerability to contaminants? Unfortunately, no one really knows yet. The seeds of the next generation have already been planted in the Arctic, but the answers could still be generations away. The circumpolar north has been transformed into an immense living laboratory, where scientists are gradually unraveling the fate of contaminants on Earth and their effects on all its inhabitants, from pole to pole. Someone once said that ecology isn't rocket science—it's much harder. Although scientists have made great progress, most answers elude them.

Tom Smith still prowls the Arctic, evaluating the health of its seals. Richard Addison retired after a career directing a team of Canadian scientists known for breaking new ground in studying contaminants in seals and whales. Derek Muir leads international efforts to document contamination throughout the Arctic, mastering the ability to spot trends, and Ross Norstrom still pushes the envelope of scientific knowledge and technology to try to unravel the effects on its wildlife. Andrew Derocher left Svalbard and returned to Canada to continue his dangerous fieldwork, warning that polar bears, jeopardized by melting ice as well as PCBs, might not survive this century. Pál Weihe and Philippe Grandjean are still testing the youngsters of the Faroe Islands—those born when the experiments began are now teenagers—and it endures as one of the longest-lasting human experiments ever conducted. They, along with Eric Dewailly, are now among the world's leading experts on the human health effects of exposure to industrial poisons and pesticides, and their findings guide world regulators in determining how much tainted fish and other seafood is safe to eat. Recognizing a global need for analyzing contaminated foods, Dewailly hopes to soon take his show on the road, developing a mobile laboratory to explore the symbiotic relationship between the oceans and human health and seek a balance between the benefits and risks of seafood around the world.

As with most environmental crises, there are no quick and easy solutions to the Arctic's dilemma. Even if the flow of all pesticides, PCBs, and other compounds is halted by every nation today, the tons already in the Arctic cannot be swept away or cleaned up. They are too ubiquitous, too persistent, too deeply embedded in the biota. Old PCBs, DDT, and other chemicals will remain there as long as it takes for nature to cleanse itself. And perhaps the most ominous discovery of all is that new chemicals are continually joining them.

Richard Addison, now retired, opens the door of a cavernous warehouse at Canada's Institute of Ocean Sciences in the seaside town of Sidney on Vancouver Island. It is 2001, twenty-nine years after that first pivotal package of blubber arrived at his lab, then in Nova Scotia. He walks along a cement floor, past piles of broken boat engines, past the carpenter's woodshop. He stops in front of a row of old, rusty white freezers. There are sixteen freezers here, and each contains a priceless collection of biological treasures collected from the Arctic. He opens them, one by one, until he locates one of his oldest keepsakes.

"Here we go," Addison says.

The freezer is crammed with about fifty yellowing plastic bags, each containing a foil-wrapped, fist-sized chunk of frozen flesh. He selects one and unwraps it. The odor of rancid meat fills the air. He reads the label, scrawled in black marker: "Male gray seal, June 1978, 100 grams of blubber."

Like Addison, many scientists have frozen a career's worth of fat and meat extracted from Arctic wildlife, as well as blood and breast milk from human beings. It was a pack of seal blubber like this that started Arctic scientists on their journey of toxic discovery.

Although long dead, these animals still have an intriguing story to tell. A new generation of biological detectives now mine these Arctic relics for pollutants—searching for chemicals no one knew were a threat when the samples were collected three decades earlier. This old, rotting blubber provides an irreplaceable archive for deciphering trends and spotting new threats. Already, these samples prove that the Arctic remains under siege, threatened by an array of new contaminants invading the northern latitudes. Addison fingers the package, wondering what other secrets this frostbitten chunk of seal blubber might reveal. He sets it down alongside the others and closes the freezer.

Frozen in time, slow to heal, the Arctic will be haunted by its toxic legacy for countless generations to come.

Chapter 3

The World's Unfortunate Laboratory

Everything about the Arctic seems eternal, as if you could return to the same spot every year and nothing would ever change, as if this landscape, so immense and forbidding, would remain eternally free of modern scars. Time, it seems, stands still here, and few people are around to mark its passage. Icebergs crafted by ancient glaciers jut out from the Arctic Ocean, standing like sentries forever guarding the frozen waters and shoreline from anything—or anyone—brave enough to trespass. But appearances are deceiving. Peer beneath the surface of the sea, as dark and opaque as black mica, hiding its secrets under a sheath of ice. Poisons lurk here, in this polar retreat. Guardians of one of Earth's last and largest wildernesses, Arctic people and animals are hundreds of miles from any significant source of pollution, living in one of the most desolate spots on the planet, yet, paradoxically, they are among the planet's most contaminated living organisms.

Two centuries ago, colonizers brought smallpox and other lethal diseases to the far North, wiping out entire communities of native people. Today, the outside world is imposing a more subtle, insidious, and intractable scourge on the Arctic. Inuit in remote areas of Greenland carry more mercury and PCBs in their bodies than anyone else on Earth, and the Canadian Inuit in nearby Nunavik and the territory of Nunavut aren't far behind. Nearly everyone tested in Greenland

and more than half of the Inuit tested in Canada exceed the concentrations of PCBs and mercury considered safe under international health guidelines, according to a 2002 AMAP report.

Dewailly, director of the Public Health Research Unit at Laval University and a leading authority on contamination of Arctic people, says it is likely that many individuals in Greenland carried such a potent brew of chemicals in the 1990s that their bodies, in technical terms, could have been declared hazardous waste. Men tested in Greenland in the early to mid-1990s had average concentrations of 15.7 parts per million of PCBs in their fat. Industrial waste that contains 50 parts per million of PCBs requires special disposal procedures because of its toxicity, and it is likely that in remote areas of Greenland, some people —including pregnant women—exceeded that level for PCBs alone, Dewailly says. Like industrial waste, their bodies don't contain just one toxic substance—they contain hundreds.

For mercury, 93 percent of women tested in east Greenland and 68 percent of women in Nunavut's Baffin region exceed the U.S. guideline designed to protect fetuses from neurological effects, compared with 16 percent for U.S. women. PCBs found in Arctic women also exceed the amounts considered hazardous to fetuses. About 95 percent of women tested in east Greenland, 73 percent on Nunavut's Baffin Island and 59 percent in Nunavik exceed Canada's "level of concern," set at 5 parts per billion in blood for pregnant women, the AMAP report says. (Fat can contain a thousand times more PCBs than blood does.) More than half of Greenlandic women are over the "action level" for PCBs, set at 100 parts per billion in blood, a level at which the Canadian government advises action to reduce exposure. Greenlanders are the only people on Earth known to exceed this standard. Although their PCB levels have been dropping in recent years, they remain inordinately high.

In Arctic Russia, meanwhile, contaminant levels appear to be rising. Long-awaited data from Russia, released in November 2004 by AMAP, the Russian Federation, and RAIPON (Association of Indigenous Peoples of the North, Siberia, and Far East), show that some of its Arctic people rank with Greenlanders as the most toxic human beings on the planet. Since the collapse of the Soviet economy, Russia's indigenous people have less access to imported foods, so they are relying more on a traditional diet of seal, walrus, and other wild animals. "In the areas of the Russian Arctic studied, practically every indigenous family consumes a significant amount of traditional food," the report says. "Families with low incomes rely to a greater extent on the local, fat-rich traditional diet. As a consequence, low-income, indigenous families are at greater risk of exposure." The scientists called the contamination from pesticides and industrial compounds "one of the most serious environmental and human health risks in the Russian Arctic."

Around two million people inhabit Siberia and the other Arctic regions of Russia, about 200,000 of them indigenous. But the 13,000 in Chukotka, across the Bering Strait from Alaska, are the main concern with respect to human health risks because they often eat the meat of marine mammals, which they ferment in soil. Not only are the animals high on the food web, which means they build up toxic compounds, but the soil used to ferment them is contaminated, too. People in coastal areas of Chukotka contain levels of two pesticides, hexachlorobenzene (HCB) and hexachlorocyclohexane (HCH), and, in some areas, PCBs and DDT, that are among the highest reported anywhere in the far North. Although many of the compounds come from within Russia, others, like the pesticide mirex, were never produced or used in either the USSR or Russia, so they clearly are moving long distances from Europe or North America.

Exposed to extreme levels of contaminants found in virtually every person tested on the planet, the Arctic's indigenous people have become the industrialized world's lab rats, the involuntary subjects of an accidental human experiment that reveals what happens when a boundless brew of chemicals builds up in an environment. "There may be only 155,000 Inuit in the entire world," says Sheila Watt-Cloutier, chair of the Inuit Circumpolar Conference, an organization that represents the Inuit of Greenland, Alaska, Canada, and Chukotka, "but the Arctic is the barometer of the health of the planet, and if the Arctic is poisoned, so are we all."

Until the 1990s, scientists searching for extreme amounts of industrial substances and pesticides examined animals and people inhabiting urban areas of North America, Europe, and Asia. But now, to find the most highly exposed people and predators, they journey to the Arctic, the world's newly discovered toxic hot spot, their preferred laboratory. Several hundred researchers, mostly from Canada and Scandinavia, are studying the effects and trends as part of AMAP, a scientific consortium created by the eight Arctic nations in 1991.

The scientists have discovered that the extraordinary loads of chemicals in Arctic people and wildlife are causing subtle injuries that jeopardize their health. These chemicals are capable of harming people and animals in ways that are hidden from the naked eye, and the impacts on living organisms are often unpredictable. They can mutate genes and damage cells, which can trigger cancer, and scramble sex hormones to render an animal's offspring hermaphroditic, feminized or even infertile. They can thin a bird's eggshells, killing its chicks; enter the brain of a human fetus and jumble its architecture; and suppress immune cells and antibodies, weakening the body's ability to fight off disease and infections. Mothers and their babies are the most vulnerable of all because the chemicals permeate the womb, moving from a mother's tissues to her fetus right at the time when her baby is growing and most susceptible to damage to its brain, reproductive organs, and immune system. Studies of infants in Greenland and Arctic Canada, exposed prenatally and through breast milk, suggest that the PCBs and other chemicals are harming children. "Subtle health effects are occurring in certain areas of the Arctic due to exposure to contaminants in traditional food, particularly for mercury and PCBs," the 2002 AMAP report says. The chemicals build up in a fetus before birth, and then the baby gets an added dose from breast milk. "The fetus and the neonate are very vulnerable to the effects of many of these contaminants during this critical period of development," the AMAP scientists considered.

So far, the effects in the Arctic seem subtle, not life-threatening. There are no dead bodies, no smoking guns. But scientists have amassed compelling evidence that Arctic inhabitants are not escaping these compounds unscathed. Children in the Arctic suffer extraordinarily high rates of infectious diseases such as ear infections that recur so often they cause permanent hearing loss. Scientists say immune suppression caused by chemical exposure could be responsible, at least in part. A study of Nunavik infants exposed in the womb to high levels of DDT and PCBs shows that they suffer more ear and respiratory infections, particularly in the first six months of life. "Nunavik has a cluster of sick babies," says Eric Dewailly. "They fill the waiting rooms of the clinics."

PCBs babies are also born with lower birth weight, and the contaminants appear to inflict neurological damage on newborns comparable in scope to that which would result if their mothers drank moderate amounts of alcohol while pregnant. Tests on Arctic and North Atlantic children show that prenatal exposure to mercury and PCBs alters their brain development, slightly reducing their intelligence and memory skills. In one new study, eleventh-month-old Nunavik babies with high PCB levels in their bodies recognized a previously seen picture 10 percent less often than Nunavik infants with low PCBs. A separate study links PCBs to impaired intelligence of children in Qaanaaq, Greenland, while similar neurological damage has been demonstrated in the children of the Faroe Islands, whose mothers ate mercury-tainted whale meat and blubber, and in children exposed to high prenatal doses of PCBs in areas outside of the Arctic, including the Great Lakes region, Germany, and the Netherlands. An AMAP human health panel in 2003 found that "at levels presently found in the Arctic, it is reasonable to conclude that the traditional diet in the Arctic contains xenobiotic substances which have a negative influence on health."

While the risks of their diet are largely unknown, the benefits are well proven, and have persisted for thousands of years, perhaps longer, back to the days when the first humans evolved. Some scientists theorize that a diet of marine organisms, like that consumed by Arctic inhabitants today, has been vital to evolution for millions of years. Life originated in the oceans half a billion years ago and all animals that roam the world today evolved from these prehistoric aquatic ancestors. More than 80 million years ago, on the bottom of what was then a shallow muddy sea, giant mollusks, the largest shellfish, thrived in the Arctic. The oldest rocks on Earth, dating back almost four billion years, were unearthed in western Greenland, and they carry the oldest signs of water and life, too. Could the genesis of life have come in the frigid waters of the Arctic? Foods from the sea are rich in nu-trients and brain-building fatty acids. Perhaps seafood is the missing link, the reason that human beings evolved with unsurpassed intelligence in the animal kingdom.

Because of this ancient dietary connection between man and the sea, the human genome may be better suited to the wild and unprocessed foods of Inuit hunters than to the modern foods of today's supermarkets. Doctors believe that the polar diet keeps hearts healthy and protects from many cancers, with benefits even greater than those of Japanese and Mediterranean diets. Unlike beef, chicken, and pork, the blubber and meat of marine mammals is low in saturated fats and high in the omega-3 polyunsaturated fatty acids that lower the risk of heart disease, the industrialized world's number one killer. The typical Inuk takes in forty times more of these beneficial fatty acids than the average American—2,000 milligrams per day compared with 50. That could explain why heart disease appears to be less prevalent in Greenland than in the rest of Europe and the United States. Dr. Gert Mulvad of the Primary Health Care Clinic in Nuuk says a seventy-year-old Inuk has coronary arteries as elastic as those of a twenty-year-old Dane who eats Western foods. Some Arctic clinics do not even stock heart medications such as nitroglycerin because heart attacks are infrequent. Although heart disease has increased with introduction of Western foods, especially among Greenlandic young people, it remains "more or less unknown," Mulvad says. Evidence that omega-3s protect the heart is emerging. In a Boston study published in 2002, men with no history of heart disease had an 81 percent reduced risk of suffering a deadly heart attack if they had high levels of omega-3s in their blood. The fatty acids seem to reduce cholesterol and prevent hardening of the arteries, as well as stabilize the heart electrically to reduce arrhythmias. They also have potent anti-inflammatory benefits. Early exposure, beginning in the womb, apparently primes the cardiovascular system at a young age, protecting from heart disease later in life.

Antioxidants in the marine mammals seem to also lower the Inuit's risk of cancer. When a team assembled by Dewailly dissected the prostate glands of Greenlandic men, "not a single cancer cell was found," he says. A diet rich in fatty acids may also reduce neurodegenerative disorders, in particular Alzheimer's disease, because they are important for brains to function normally. Much of the brain consists of fatty acids, and neurologists already know that they are essential for brains to develop normally in the womb. Passed to a fetus or infant through the placenta or breast milk, they promote brain development. Ironically, those are the same functions of child development that appear to be impaired by two of the main contaminants in Arctic foods, mercury and PCBs. As a result, the contamination may be nullifying the food's benefits to the fetal brain. Public health officials are torn between encouraging the Inuit to keep eating their traditional foods and advising them to reduce their consumption. Doctors fear that the Inuit will switch to processed foods loaded with carbohydrates and sugar if they are told to limit marine foods. "The level of contamination is very high in Greenland but there's a lot of Western food that is worse than the poisons," Mulvad says.

Traditionally, their marine diet has made the Inuit among the world's healthiest people. Minus the contaminants, this diet is arguably the most nutritious in the world—loaded with vitamins, minerals, antioxidants and protein, as well as the fatty acids. Dewailly says eating marine mammals and fish is "like getting a huge vaccine in your food." Beluga meat contains ten times more iron than beef, five times more vitamin A, and 50 percent more protein. Six ounces of narwhal skin contains 63 mg of vitamin C, about the same as a glass of orange juice or cup of strawberries. Despite a lack of fresh fruit and vegetables, Arctic inhabitants get virtually every nutrient their bodies need—including riboflavin, thiamin, and niacin, and the antioxidant vitamins A, C, E, and selenium—from their native foods. About half the iron consumed by schoolchildren on Baffin Island comes from the meat of marine mammals.

If Arctic people are forced to stop eating their native foods because of contaminants, vitamin deficiencies, malnutrition, and disease—what nineteenth-century explorers called the Arctic scourge—would climb. The scourge that struck these early expedition teams started with blackened gums and loose teeth. Before long, their hair fell out and their joints turned stiff and sore. They rapidly became invalids, and many died a painful death. Untold numbers of men from European and American teams exploring polar waters for the Northwest Passage or North Pole died of scurvy, the Arctic scourge. While the Inuit they encountered on their journeys seemed immune, every expedition to the Arctic was ravaged by the disease. In their voyages through Baffin Bay, members of the British Royal Navy—the first white men to winter in the Arctic—shunned Inuit foods and instead ate salt meat they had brought along. Their commander, Lieutenant William Edward Parry, was forced, twice, to return home to England, his crew decimated by disease. The explorers didn't know that they were surrounded by the only "medicine" they needed: the meat and blubber of seals and whales, a rich source of essential nutrients, including vitamin C, which prevents scurvy. Medical experts also extol the nutritional benefits of breast milk, even in the Arctic, where it carries such a large load of contaminants. The greatest threat that chemicals pose comes in utero, so the amounts passed to the newborn through the milk are considered less harmful. Few foods are as nutritious as mother's milk, so its benefits are thought to outweigh its risks.

If the Arctic is suffering from contaminants, what about the rest of us? What does this say about the animals and people where pesticides are sprayed, where power plants spew fumes, where plastics leak industrial compounds? If high levels of contaminants were found in chicken or hamburgers, rather than seal, narwhal, and beluga, it is entirely likely that the Western world would not tolerate it. But excessive levels of contamination are not limited to the Arctic. People in an array of locations in both the Northern and Southern Hemispheres—the Great Lakes, the Baltic countries, Southeast Asia, Africa—are highly exposed to chemicals such as DDT and mercury, especially in seafood-eating cultures. In the United States, one of every six babies—about 630,000 a year—is born to a mother carrying more mercury in her body than considered safe under federal guidelines, and the source is fish, particularly large, predatory species like swordfish, shark, and albacore. When it comes to the newer brominated flame retardants, the highest global concentrations are in U.S. women. Yet in the Western world, most people can freely make choices in their diet to limit their risks, while the Inuit face an unfathomable dilemma: either reduce their reliance on the only foods that sustain them, particularly seal meat and whale blubber, or expose their children to risks. For inhabitants of the polar world, there is no real choice. To survive—as individuals and as a society—they must keep hunting marine animals.

The pollution is a social injustice because wealthy industrial regions are imperiling the resource-based economy of distant lands. "All over the Arctic, environmental pollution is restricting the achievement of sustainable food security," says a report coauthored by Gerard Duhaime, an economic sociologist at Laval University. Lars-Otto Reiersen, AMAP's executive secretary, says it is particularly egregious because Arctic people do not use or benefit from the chemicals and pesticides. "We have a moral duty to do something," Reiersen says. "We cannot send dirt to our neighbors and close our eyes. It is clear that the Arctic acts as a global watchdog. When changes can be documented in these remote areas, there also are effects further south, closer to the sources of the problems."

Anthropologists say the contamination jeopardizes more than health; it threatens sweeping societal changes in the Arctic akin to cultural genocide. Efforts to alter Inuit diets can unwittingly trigger permanent cultural changes. Hunting embodies everything in the Inuit's 4,500-year-old society: their language, their art, their clothing, their legends, their celebrations, their community ties, their economy, their spirituality. "It's not just food on a plate," Watt-Cloutier says. "It's a way of life."

The Arctic's human inhabitants share their place at the top of the world with other meat-eaters. The larger, older, and more voracious a creature is, the more contaminants it carries. And few animals on Earth live as long or eat as much as a polar bear.

Chapter 4

Plight of the Ice Bear:
Top of the World, Top
of the Food Web

Born at Christmastime, cradled in pure white snow, two newborns are sleeping and suckling, swaddled in the protection of one of the fiercest creatures on Earth. The brothers were born blind, toothless, a pound apiece, as feeble as kittens. For four months they nestle in a den carved by their mother on the snowy banks of a frozen sea, where they gorge themselves on rich, fatty milk, doubling their weight every few weeks. As spring arrives, darkness gives way to days and nights that shine with eternal light, and the cubs step outside for the first time, scampering along the sea ice trying to keep up with their mother. If they beat the odds and survive their fragile first year, they will reign as king of the Arctic, Norway's *isbjorn,* the ice bear.

Spring has arrived at this polar bear nursery on the Norwegian archipelago of Svalbard. Carved by harsh winds and ancient glaciers, closer to the North Pole than to Oslo, Svalbard is one of Earth's last true wildernesses, a place so far north that compasses are useless. Gashed by deep fjords and spiked by jagged peaks, Svalbard (in English, "the cold coast") is an icy, roadless archipelago four times the size of the Hawaiian Islands. It was discovered by Vikings in 1194, who mistook it for part of Greenland. In fact, it is closer to Greenland than to Norway, since the Norwegian mainland is 400 miles away. In the northern reaches of Spitsbergen, its largest island, the sea never melts. Almost two-thirds of the islands are covered with solid, year-round ice. In some spots, the permafrost extends down a quarter of a mile. Svalbard, like the rest of the Arctic, is an ice desert, almost as dry as the Mojave, with only a few inches of precipitation per year.

It is early evening, nine days after the return of the midnight sun in April, 2002. Brilliant cobalt-blue seawater is shattering the islands' icy shield, splitting the frozen fjords into patches, like frosty white lily pads floating on a pond. Plankton is beginning to bloom, providing a feast that draws hordes of fish, migratory birds, and seals. Polar bears are the largest land carnivores on Earth today, but they are truly marine animals. The sea ice is their realm, and they move with it as it recedes each spring.

On overcast days like this, it is hard to tell where the ice ends and the clouds begin. Shades of white blend seamlessly, and the horizon is lost. The land-fast sea ice, less than a year old but already solid, looks as taut as a bedsheet in some spots, as billowy as a down comforter in others. This vast white prairie is a favorite spot for polar bears to raise their cubs. On this unseasonably warm day, a mother bear has abandoned her winter den to begin the tutoring of her two sons. They are walking on the finger of a frozen fjord named *Woodfjorden,* almost 80° north latitude, six hundred miles from the North Pole. These bears are the undisputed king of the Arctic, which derives its name from *arctos,* Greek for bear. But from the moment of birth—even conception—they struggle against the odds. Most newborns die of starvation even under the best natural conditions. Yet it is an unnatural threat—a man-made one—that is intruding on the desolate wilderness of the High Arctic and jeopardizing the ice bears' survival. At the top of the world and the top of the food web, polar bears are born at the wrong place at the wrong time. Despite living in the remote reaches of the High Arctic, bears inhabiting Svalbard and Russia's nearby Franz Josef Land and Kara Sea carry extraordinary loads of industrial chemicals, a dose more toxic than that found in most other living animals anywhere. Masses of air and ocean water pass by Svalbard, dropping off pollutants that hitchhiked from Europe, Russia, and the east coast of

North America. Because atmospheric patterns favor transport to the European archipelago rather than the western Arctic, Svalbard's air contains the highest concentrations of PCBs found in the Arctic, so its bears are more highly exposed than their counterparts elsewhere. On average, Svalbard bears are twelve times more contaminated than Alaskan bears.

Before they even leave the safety of their dens, these cubs are poisoned by their mother's milk; upon birth, they harbor more industrial pollutants in their bodies than most other creatures on Earth. When cubs feast on their mother's milk, they feast on her past. A mother polar bear eats voluminous amounts of seal blubber, amassing a lifetime of toxic chemicals in her own fat. There it remains until she unwittingly bequeaths it to the next and final step up the food web: her cubs. After just a few weeks of drinking their mother's milk, which is one-third fat, the cubs carry higher concentrations of PCBs than their mothers and billions of times more than the waters of the Arctic Ocean. About 2,000 bears, perhaps 10 percent of the world's population, inhabit these islands, and scientists are undertaking perilous expeditions there to chronicle their fate.

The mother bear lumbers along, hunting her favorite prey of ringed seals, leaving a zigzagged path of twelve-inch-wide craters followed by the smaller pawprints of her cubs. A few miles away, from the front seat of a helicopter, Andrew Derocher of the Norwegian Polar Institute has spotted the bear family's fresh trail. Pilot Oddvar Instanes skillfully loops, spins, and straddles the tracks, following their erratic path for several miles. Lounging by a hole in the ice, a seal looks up, puzzled by the helicopter's noisy antics.

"She's running here," Derocher says, pointing to a row of tracks at the edge of a craggy glacier. "I think she's ahead of us here somewhere."

It is Derocher's seventh season tracking Svalbard's bears, monitoring their health and testing them for contaminants. Only about a dozen people on Earth know how to find and catch a polar bear, and

Derocher is one of the best. In twenty years of research in the remote reaches of the Canadian and Norwegian Arctic, he has caught about four thousand. Because their fur is pigment-free, translucent as ice, and the hollow cores of the shafts reflect light, it is easier to spot their tracks than to spot the bears themselves. Their tracks crisscross beneath the helicopter like mazes, old ones merging with new ones, and Derocher manages to untangle them from above. A geologist once advised him to switch to studying glaciers—they move a lot more slowly than bears. But after all these years, Derocher knows where to find them. He eyes the edges between thick and thin ice and the rough fronts of glaciers, ideal places to detect fresh tracks. The helicopter hovers 300 feet above the surface—close enough for Derocher to distinguish a bear from a fox but high enough for his eyes to sweep miles of ice at a time. Even from that distance, Derocher can tell that these tracks were made by a mother and two new cubs. Derocher has picked up their trail, and the mother and cubs are now walking right below their chopper. In the backseat, Magnus Andersen, his young, blond Norwegian colleague, fills a syringe with a colorless liquid, the same tranquilizer that veterinarians use, in smaller doses, to anesthetize dogs and cats for surgery. He injects the drug into a dart and screws it onto a modified shotgun equipped with a .22-caliber blank.

Oddvar dips the chopper to about six feet over the mother's head, so close they can see the coarse hair on her back blowing in the wind. Silently, Andersen kneels on one leg and opens the door. A freezing blast of air slaps him in the face. The blades whip up a frenetic whirlwind of snow, masking his view. The mother bear starts to run, and the helicopter, its engine roaring, follows the bear, spinning in 360-degree circles perilously close to the ground and turning sideways to give Andersen a good, clear shot. Spinning, swirling, spinning, swirling, spinning, swirling, again and again. Adrenaline pumping, Andersen leans his head and shoulders out the open door, attached by a thin, green mountaineer's cord. He takes aim and fires. A muffled, dull thud. The smell of gunpowder wafts through the door.

"OK," Andersen says. A dart sticks from the bear's rump, so Andersen is pleased. Precision is important. If the dart had hit her in the chest, it would have killed her, as sure as a bullet. From above, Andersen and Derocher follow her path, watching to make sure she doesn't stray onto ice that is too steep or cratered for the chopper to land on. They must avoid chasing her and the cubs; bears running long distances can easily overheat and die of heatstroke. Within minutes, she starts to wobble but she still isn't going down. "Take her in the neck if you have to," Derocher says. Andersen readies another syringe and fires the shotgun, hitting her in the rump again. This time, she lies down on her stomach, eyes open, her body perfectly still, one giant paw splayed back. The cubs nuzzle her, trying to awaken her, then curl up beside her. Peering out over their mother's back, they are wide-eyed and curious, still panting from the run, as the helicopter lands and Oddvar cuts its engine. The doors open to silence.

Derocher and Andersen cautiously approach on foot, their boots crunching in the crusty snow. The two men circle slowly around the bears. At six feet three inches tall and 225 pounds, Derocher's towering height, jet-black hair, and full beard give him the aura of a big black bear. But this mother polar bear is twice his weight, and a male weighs much more, almost a ton. He knows polar bears enough to fear them, and he and Andersen always carry loaded .44-magnum pistols holstered on their waists. A few years earlier, two young tourists in the Svalbard town of Longyearbyen were mauled to death. Now, as soon as visitors set foot on the islands, they are handed a pamphlet with a photo of two bears ripping apart a carcass—a seal, presumably. Its entrails are exposed in a bloody pulp and bold scarlet letters warn: "TAKE THE POLAR BEAR DANGER SERIOUSLY!" Derocher never forgets that advice. Even though he has been hurt only once—when a cub swung around and bit his finger, leaving a small, shallow cut—he doesn't like being on the bears' turf. Bears and people inhabit Svalbard in roughly equal numbers, but the bears definitely have the advantage in this icy realm. "It's never the bear we're drugging that's dangerous," he says in a Canadian accent that sounds a bit Irish in its rustic lilt. "It's always the bear you don't see. If there's one bear, there's two bears. If there's two bears, there's three." Once while he and his team worked on a bear, two others approached in the distance and they raced to the helicopter to avoid being attacked. "I've never shot a bear," Derocher says. "If you do that, you've made a mistake. A big one." After all, their mission here is to protect the bears, not harm them.

Gloveless, Derocher strokes one cub's creamy-white fur and Andersen holds out a finger for the other to sniff and lick. The four-month-old cubs are as cuddly as their mother is deadly. Their fur is soft while their mother's is stiff and bristly, coated with ice crystals. At about fifty pounds apiece, the cubs are the size of Derocher's daughter, who was six years old at the time. These are the first humans these cubs have seen, and perhaps the last. Andersen gently loops ropes around their necks and tethers them to their mother. Without her, they would die.

Andersen checks the mother's ear for an identifying tag. "She was caught once before," he says.

"When?" Derocher asks.

"In 1994."

Derocher sets down his black toolbox, removes some dentist's pliers and pries open the bear's massive jaw. Leaning inside her gaping mouth, he deftly extracts a tooth from behind the main canines, a useless premolar the size of a cribbage peg that he will use to confirm her age. She is around fifteen years old, and Derocher wonders if this will be her last set of cubs. Older females like her are rarely seen denning in Svalbard, even though polar bears live as long as twenty-eight years in the wild. Andersen works on her other end, trimming some fur with a scissors and using a biopsy tool to slice a quarter-inch-diameter plug of blubber from her rump. Then he pricks a vein in the thigh of her hind leg with a needle, quickly siphoning a tube of blood. Derocher rolls her over and checks her reproductive organs. They are normal. Together, the two scientists stretch a rope over the mother to measure her girth and length, which is then used to extrapolate her weight. Derocher jots the numbers down in a notebook using pencil, since he learned the hard way years ago that pen ink freezes. He and Andersen always work with bare hands, whatever the weather. On this April day it is extraordinarily warm for spring in Svalbard, right around the freezing mark, but three days earlier, they worked on bears in temperatures of minus 2 degrees Fahrenheit.

It is time to work on the cubs, but first Andersen and Derocher inject each with tranquilizer so they can take measurements and blood samples and weigh them on a hand scale. Together, they hoist the first cub up—it weighs in at fifty-four pounds. The second is forty-five pounds. With a clamp, Derocher attaches a tag marked with identifying numbers to the cubs' ears. Drops of blood drizzle onto the snow. Derocher kneels beside the mother and milks her like a cow to sample the creamy liquid she is feeding her sons. The milk and fat will be analyzed at a lab for a suite of chemicals, and their blood will be checked for various indicators of their health, from sex hormones to immune cells. Then Derocher lifts her giant head and puts her lolling tongue back in her mouth. Oddvar paints a big brown X on her rump, signaling that she shouldn't be bothered again this year. The cubs are left snoring, all eight paws splayed out on the snow. The threesome will sleep for two hours, then shake off the drowsiness and continue on their way. The scientists pack up their toolbox and silently walk back to the helicopter. It has been forty minutes since they landed.

Capturing polar bears can be dangerous, for both man and bear. But scientists say it is critical for understanding how wild animals are faring, and what chemicals they carry in their bodies. "Otherwise," Derocher says, "we would blindly stumble into extinction. My job is to make sure polar bears are around for the long term. The Arctic without polar bears would be like the plains without buffalo." Every April for seven years, Derocher has left his wife and children for a month of field-work in this frigid laboratory, capturing between sixty and one hundred of Svalbard's bears as the mothers and cubs leave their dens. He is just shy of forty-two years old on this trip, yet he already has noticed that each spring it takes longer to warm his stiffened joints upon returning home to the polar institute in Tromso, on Norway's Arctic mainland. Usually by his age, the frigid weather wears down Arctic scientists and they leave the fieldwork to their younger and more resilient graduate students. Yet Derocher can't imagine stopping or even slowing down. His heroes are the nineteenth-century polar explorers, particularly the Norwegian Fridtjof Nansen, who set out on uncharted ice by ship, sledge, and skis, surviving for years at a time with few provisions. "They went out in the great unknown, making magnificent discoveries. It was a magical time," Derocher says. He dismisses contemporary polar expeditions, with their freeze-dried food, vitamin C tablets, satellite phones, and global positioning systems. There is considerable adventure to his own work, but he rejects the notion of any comparison with the early explorers. He is a scientist, not an adventurer, and frankly, he admits, he hates being cold. "I don't think I would last a month out here," he says. "Not unless I had my Gore-Tex and fleece and high-powered rifle." Andersen remembers the first time he leaned out the helicopter to shoot a bear with tranquilizer. It was terrifying, but it wasn't the fall he feared; it was the bears. He wants to make absolutely sure they are sound asleep before he walks among them.

Weather is the biggest threat to polar bear scientists, who must maneuver Svalbard's dangerous mountain passes and glaciers. Forecasts here are mostly unreliable, and when bad weather sets in, sometimes without even a minute's warning, Derocher and his team can be stranded on the ice for days. Or worse. On a spring day in 2000, Malcolm Ramsey and Stuart Innis, two of his highly experienced colleagues in Canada, were done for the day, on their way back to base, when they were caught in a whiteout—clouds descending with no warning. Their helicopter crashed into a glacier. Now, when caught in whiteouts, Derocher and his crew throw black garbage bags filled with rocks out the window. Sometimes it's the only way to know which way is up in Svalbard. He always packs three radios and three global positioning devices, and when in doubt of approaching weather conditions, he directs the pilot to land on the ice and wait it out. "There's no data worth giving up your life for," Derocher says. "Making a wrong decision can be fatal. You don't want to do that too often."

The helicopter lifts off, headed north between snow-draped peaks. Within ten minutes, next to a rambling crack in the sea ice, Derocher spots more tracks—this time, a mother and two plump yearlings. Andersen fills another syringe and rests the shotgun on his leg.

Born in Vancouver but claustrophobic in cities, Derocher is guided by his own internal compass that steers him north, far north, whenever he craves serenity. As a child, he moved to Vancouver's north shore and spent his free time exploring the banks of the Fraser River and poring over library books about natural history and wildlife, no animal in particular, "just about anything alive," he says. He was seven years old the first time he saw a bear in the wild. He and his father were driving through a provincial park on a camping trip when they saw three black bear cubs romping on a grassy knoll. He begged to play with them, but his father refused, warning that the mother bear was there somewhere, ready to attack. Derocher was intrigued by the combination of their fearsome image and mysterious nature, and when he grew up and studied zoology at the University of British Columbia, he decided to concentrate on large mammals, focusing first on the forest habitat of grizzlies and black bears. "To this day, I don't trust bears. I don't trust them for a minute," Derocher says.

In 1985, when Derocher ventured as a young biologist into the Canadian Arctic for the first time, it struck him as a barren wasteland compared with the lush forests he was used to. Then his mentor, renowned Canadian polar bear biologist Ian Stirling, dropped a hydrophone into the frozen ocean. When Derocher heard the secret undersea symphony—whales singing, glaciers crashing, seals grunting—and saw bloodstains on the ice from polar bears feasting on seals, he realized this was no sterile wasteland. He was instantly hooked. The High Arctic, he says, "is the end of human civilization, the start of the last true wilderness." He wants to delve into how polar animals survive, how they adapt to their environment. To him, it has always seemed like black magic to travel to remote latitudes—so far north that compasses fail—yet find your way back again. In the early days of his research, the helicopter would drop Derocher and the other scientists off, leaving them on the ice. He grew to love the elegance of emptiness, the splendor of a place so raw. "I find the Arctic an incredibly serene environment. Far off on the sea ice there is an immense sense of peace and remoteness that you can't find in many places in the world anymore," Derocher says. "There is a sense of humility, I think. It really is an intense wilderness that is hard to match in a terrestrial environment. It may be a little like being out in a sailboat in the middle of the Atlantic. For humans, most of us aren't creatures of the sea ice. We don't have a strong mental connection with those environments, so they feel very foreign to us."

Svalbard, he thought, would be the perfect laboratory for studying polar bears in their natural environment, pure and pristine, with no hunters. "One of the reasons I went to Norway is they didn't have a harvest. They hadn't had one for twenty years," he says. But when he arrived in 1996 for his first year of fieldwork there, it somehow felt like a harvested population. Something, he knew, was wrong. Derocher's image of the archipelago was shattered when an array of toxic chemicals were found in the polar bears' bodies in the mid- to late 1990s. "It is no longer a natural situation. Polar bears carry a huge variety of pollutants," Derocher says. "Try and convince me that they have no impact on the animal's physiology."

Polar bears are exposed to hundreds of contaminants, including DDT, but, for some reason, they are able to metabolize most of them, purging them from their bodies. The exception is PCBs, which build up in their tissues. In Svalbard bears, PCB levels peaked at 80 parts per million (ppm)—which Norwegian scientists consider an alarmingly high amount—and averaged about 30 ppm between 1987 and 1995. To the east, around Russia's Franz Josef Land and Kara Sea, polar bears have even higher PCB levels—twice as much, with an average estimated at 60 ppm in the 1990s. The region's other top predators—Arctic foxes and glaucous gulls—are highly contaminated, too.

Although the impacts of contaminants on the Svalbard bear population are not well understood, there are signs that they could be causing mortality and reproductive impairment of females and lowering the survival rates of cubs, Derocher and colleagues reported in the journal *Science of the Total Environment* in 2003. Derocher suspects that the chemicals are reducing the bear population by 2 to 4 percent per year, an effect similar in magnitude to hunting when it was allowed on Svalbard. "Everything indicates that the polar bears are being affected by these contaminants," said Geir Gabrielsen, the Norwegian Polar Institute's director of ecotoxicology. "There are so many indications that there are population effects." Yet proof is elusive, because today's doses cause symptoms that are subtle and difficult to diagnose in wildlife. "These contaminants aren't a sledgehammer," said Swedish researcher Cynthia de Wit, who authored an AMAP report on effects on wildlife. "They might just be a regular hammer."

PCBs and other organochlorines are endocrine disruptors, which means they can alter hormones of a fetus, and in Svalbard's bears, they have been linked to a long series of hormone-related biochemical changes. Norwegian scientists who examined the blood taken from the bears found altered testosterone and progesterone, which are sex hormones important to reproduction, as well as suppressed immune cells and antibodies. They also found reductions in thyroid hormones, which control how the brain develops; retinol, which regulates growth of skin, bones, and reproductive organs; and cortisol, a steroid hormone important for managing stress, blood pressure, and other vital functions. Even the mineral composition of the bears' bones seems to be altered by the chemicals. Scientist Christian Sonne linked osteoporosis in bears with high levels of PCBs in a 2004 report.

In essence, virtually every inner part of the bear's body is under assault. However, what scientists aren't sure about yet is what these biological changes mean to the health of individual bears or to the whole population. Some bears carry such high levels of PCBs in their blood that their immune systems might not be able to muster the strength they need to fight off viruses, bacteria, and parasites. Producing a flood of antibodies to fight off viruses and infections is critical for an animal's survival but when Svalbard polar bears are exposed to a flu virus in experiments, they cannot manufacture as many antibodies as Canadian bears that have far less PCBs, according to studies by the Norwegian School of Veterinary Science. No mass die-offs of wildlife have ever been documented in the Arctic, but in Europe's North Sea, two distemper epidemics, in 2002 and 1988, wiped out an estimated 38,000 seals. Experts specializing in immunotoxicology say the seals' high PCB levels exacerbated the scope of the epidemic.

The altered testosterone and progesterone hormones could be reducing the bears' fertility and perhaps even mixing up their reproductive organs at birth, causing the pseudohermaphrodites, bears born with both female and male reproductive tissue, that Derocher found on Svalbard. Of every hundred bears Derocher has captured, three or four have female and partial male genitalia, although they remain fertile and able to produce cubs. Scientists have found animals with similar mixed-up sexuality in other areas polluted with pesticides and other contaminants that mimic estrogen or block testosterone, including alligators in Florida lakes and fish in British streams.

Yet the pseudohermaphroditic bears could be a natural phenomenon. Sonne and other scientists from Denmark's National Environmental Research Institute, who examined an aging, unusual bear in Greenland, theorized in a 2005 report that the ones that seem to have small penises actually have enlarged clitorises due to rough sex, as sometimes seen in dogs. They believe the condition has nothing to do with chemical exposure. Derocher, however, remains unconvinced. "I find the issue of rough sex harder to believe than the pollution issue," he says. "I think there are reasons to still suspect pollutants." For one, he says, he found twin yearlings with the condition. "The bottom line is that we don't know why [this condition] exists," he says. "It certainly seems a lot more common in Svalbard than anywhere else in the world. Is it coincidence that Greenland, also very polluted, has a case as well? I guess with time the issue may be clearer."

Ross Norstrom of the Canadian Wildlife Service, one of the world's leading polar bear experts, worries most about the cubs. Because they weigh a mere one pound at birth, development of their immune and reproductive systems occurs during their first few months, right when they are hit with a huge blast of PCBs from their mothers' milk. The cubs of mothers with high levels of PCBs in their milk are more likely to die during their first year than cubs of mothers with low PCBs, according to research on Canadian bears. Denning mothers who lost their cubs had three times higher concentrations than mothers whose cubs survived. Janneche Utne Skaare, of Norway's National Veterinary Institute, says Svalbard's bears are denning more often than bears elsewhere, closer to every two years than three years, a possible sign that their cubs are dying early. "Cub survival in the Svalbard population is lower than in other populations, and the high PCB concentrations found in cubs compared with older bears could be a possible explanation," an AMAP report says.

Derocher and other scientists, hampered by the extreme expense and hardships of polar expeditions, cannot say definitively whether the pollution problems facing Svalbard's polar bears are reaching a crisis. There is no bear census. Scientists can't check the birth records of bears. Instead, they only get brief glimpses of their lives. From fifty to one hundred bears out of an estimated two thousand on Svalbard are caught each year, and the scientists share only about half an hour with each one of them. They know little, if anything, about where a bear has been and where it is going. They can't monitor which ones die, or what they die from, and they aren't even certain whether their numbers are rising or falling on Svalbard. What they do know, they glean from their biological detective work—the testing of the bears' blood, fat, and milk. Lack of insight and lack of funding are what prevents scientists from gaining a better understanding of the polar bears' plight. Their status, Derocher says with a shrug, "is at best unknown. We are still in the very steep part of the learning curve about the effects of toxic chemicals on wildlife."

There are no practical ways to cleanse Svalbard of its contaminants. PCBs in the bears are declining now, but the levels remain excessive a quarter-century after the compounds were banned in most industrialized nations. Making matters worse, polar bears face the threat of melting ice. Climate change has dramatically thinned sea ice, wiping out many of their hunting grounds. If the rate of ice loss continues unchecked, Derocher predicts polar bears will become extinct in some regions within one hundred years.

A storm is approaching Longyearbyen as the team climbs into the chopper at 8:40 A.M. "If it's too windy, we'll come back," Derocher says. Instanes, a map spread across his lap, radios to a dispatcher that he is headed to the eastern side of Spitsbergen, four people aboard. The blades begin to spin, and he pilots the chopper into the headwinds. Derocher holds up his binoculars, training them on the ice below. Nothing is stirring.

"What do you think?" Derocher asks. "I don't think it looks very good myself," Instanes says. The snow is too hard from recent rains and the winds too strong to find tracks.

A few minutes later, Derocher asks the pilot: "What are you thinking now?"

"Let's go over here," Instanes responds, pointing.

"How's the wind?" Derocher asks.

"It's OK. It was about twenty knots at the beginning. But it will be OK." They can't see the winds, except for an occasional flurry of snow flying past the windshield, but Instanes knows that once they see them, it will be too late to land safely. The chopper is only a few yards off the ground, and sudden gusts can thrust it to the ground, tail first, then flip it forward.

A few minutes later, a half-mile into the trip, Instanes announces, "Now I can see the wind. We've got strong winds here. Very strong.

I think we should head home. Do you want to go a little farther?" "My concern is we're going to be stuck on this side," Derocher answers. "We're just looking for trouble going down in that stuff." He is most worried about the light—it is flat, with poor depth perception, making tracking dangerous.

The crew, dejected, heads back to the polar institute, touching down in Longyearbyen at 9:15 A.M. "Yes, well. Good time for a coffee," Derocher says.

Polar bear research is a constant balancing act: Waiting for good weather wastes money, but impatience can cost you and your crew their lives. Derocher sometimes feels as though he is sitting in the helicopter with a big bag of money on his lap, throwing thousand-dollar bills out the window. The helicopter and pilot cost more than $2,000 (U.S.) per hour, and government funding is always unpredictable. He was allotted only sixty hours of flying time this particular year. Although costs had doubled since he arrived seven years earlier, the Norwegian government's funding had stayed the same. Costly decisions about when to fly must be made instantaneously, on the ground and in the air. Is bad weather moving in? Is there enough fuel to get back? Is that ice safe to land a bear? Sometimes he just has to rely on his instincts. Safety is number one. But finding bears is number two. With each wasted day, Derocher grows impatient. The money is about to run out for the season. It's been a strange season for the team of scientists. A good day here and there, followed by many more days waiting for weather to clear. A half-dozen bears caught in one day, none the next. Derocher, Andersen, and Instanes had just camped for three weeks on Hopen, an island east of Spitsbergen so remote that they had only a radioman, a weatherman, and five huskies for company. On Hopen, they usually have access to a few hundred bears. Yet the conditions were so dismal—smashed-up ice, stormy weather—that they caught only thirty-four. So they pulled out a few days early and headed here, to Spitsbergen, where so far, their luck has been better but not great. They are now up to forty-eight bears for the season, but Derocher is eager to break one hundred and they have only five days left before their helicopter contract expires. On a good day, he can find twenty bears. But the weather—and the bears—must cooperate.

Derocher pulls off his boots and jacket, padding around in his wool socks. He flips on a computer to check a weather satellite on the Internet, and contemplates heading north because winds are stiller there. He knows many bears are there, so he is tempted. Derocher huddles over the computer with Andersen and a geologist, looking at the satellite images. Finally, he says with a sigh, "I think we're drinking coffee and checking e-mail today."

"Sometimes you hope the weather gets really bad," he says, "because then it can only get better." Derocher left his family in Tromso nearly a full month earlier, in late March. When he called home, his wife told him that it was 48 degrees Fahrenheit at home—about fifty degrees warmer than in Svalbard—and flowers were blooming. The end of polar winter is a joyous time in Tromso, the Arctic's most cosmopolitan town, but Derocher never gets to enjoy it because he rarely returns home before early May.

He seeks out Instanes for advice, then returns, their decision made. "Let's go," Derocher says.

The winds are still howling as they set out again at 10:20 A.M., this time heading north over the fjord, hugging the coast. Flocks of seabirds hover below them as they fly over a sea so deep and dark it looks like a sheath of black stone. "I don't know what it will be like up here," Derocher says. "You never know." His voice trails off. Instanes says he is looking forward to seeing the northern fjords. "I'd rather see some bears," Derocher says with a laugh.

The sky looks clear to the north, and they sail over the white caps of the low-slung mountains, toward *Woodfjorden,* seventy-five miles away. In the bright sunlight, the ice sparkles below, as if it were dancing with fireflies. The frozen sea is cracked and etched by the wind, like skid marks on a runway. But there are no visible bear tracks. This place seems untouched by man, untouched by everything. Derocher looks down at the cracks and drifts, struggling to "read" the ice.

Suddenly, he thinks he sees tracks. Instanes makes a quick turn, then another, gliding over the tracks. "Big skinny male below us," Derocher announces. The bear looks up and runs, confused. "He's very skinny. Holy Christ," Derocher says.

Andersen readies a syringe, then fires, hitting the bear in the rump. The bear slows and stumbles to the ground, his head between his paws. The chopper lands.

"He's one of the oldest bears I've ever seen," Derocher says upon examining him. "Not much more than skin and bones. He's probably here to die." The bear is around twenty-four years old and weighs about 450 pounds, but a bear that size should be two or three times heavier. His bones poke through his fur. Derocher examines his ear, looking for an identification tag, but finds none, which means he has never been caught before. The ear is torn, almost gone, probably from combat with other bears. He grunts occasionally as they take samples of his blood and sparse fat, then mark his rump with an X and clip a tag to his ear.

"So long, old boy," Derocher says. They leave him sleeping, spread out on the ice like a living, breathing bear rug. The chopper blades spin.

Two reindeer dash across the ice. Then an Arctic fox, a rare blue one, scurries below, trying to escape the strange, noisy creature hovering above. Derocher spots more bear tracks, a mother with cubs. He smiles. The cubs are big and plump. "We broke fifty!" he exclaims.

When they finish work on the threesome, clouds are closing in on them. Derocher eyes the sky. "We'd better start thinking about making our way home," he says. Instanes assures him that the conditions are safe. They fly to a refueling station at Ny-Alesund—little more than five barrels of fuel sitting on the ice, a few cabins, and a snowmobile— then circle back over the old bear. "He's fine," Derocher says. "He just needs a little more time to wake up." They soon spot another male, this one so big and powerful, close to nine hundred pounds, that Andersen has to shoot it with two potent tranquilizer darts before it goes down. It's the fifth bear of the day, the fifty-third of the season.

Low clouds are descending and they decide to wrap things up for the night, to go back and prepare their blood and fat samples. Derocher is pleased. A successful trip, he says, is getting a bear on the way out and a bear on the way in—and a bunch in between. It was a lucky day— although a risky one too.

The next day is Friday, and they wake up to a blustery morning. Derocher and Andersen move slowly at the barracks, fixing a cup of instant coffee before heading into the institute office at 10 A.M. En route to town, they see some tourists walking along the road. "Polar bear food," Derocher says. He's quiet for a moment. "Bears do come along here," he says, and he and Andersen recall how two girls were mauled to death near here. When they arrive at the office, winds are rattling the windows and snow is blowing almost horizontally. Outside, pass-ersby are slipping and sliding as they walk across the frozen parking lot. A child in a red snowsuit is pushing a bicycle on the ice, making little progress in the headwinds. A mother holds tightly onto a stroller. Derocher spends the whole day—and the next—reading reports, checking e-mail, feeling restless and agitated. Sunday, though, brings a break in the weather and the crew's best day yet.

Just before 9 P.M., Instanes turns the helicopter south, back toward their base in Longyearbyen. Bad weather is closing in on them. Clouds to the north are threatening a whiteout but, miraculously, a perfect path of crystalline skies has opened to the south. The landscape below looks almost voluptuous now. Curvaceous, rounded peaks are bathed in soft light, awash in hues of icy blue and frosty white. Svalbard seems less threatening, as if it could enfold the team in a warm embrace. A cir-cular prism of pastel colors shimmers in the sky beside their helicop-ter, tracking their every movement. It is merely their reflection, but it hovers like a beacon guiding them home. The three men are glowing with their own internal light, the satisfaction—and relief—of knowing they are headed back for a hot dinner and warm bed before they wake up and head out again. They captured six bears in six hours, on a single tank of fuel, and all are safe, men and bears. Andersen unwraps a bar of chocolate he had saved for just such a moment, and shares it with Derocher and Instanes. Later, they will celebrate at Longyearbyen's best restaurant, clinking glasses of wine and toasting the day's work.

As the helicopter heads south, flying low and fast, Derocher peers out the helicopter window and snaps a photograph, allowing himself to relax enough to enjoy the view for the first time since he started the season's work a month earlier. "Boy, it's pretty when the light is like this," he says. Instanes nods. Derocher doesn't mention it, but this is his last foray into Svalbard before leaving the Norwegian Polar Institute and heading back to his alma mater, the University of Alberta, to start new bear research in Canada. Andersen and a British scientist will carry on the work for him. Derocher will leave before the next generation of bears is born with contaminants already building up in their bodies, before he can glean answers to the most disturbing questions haunting Svalbard's *isbjorn*. His seven years there weren't enough to illuminate the ecological mysteries of Svalbard and predict, with any certainty at all, the future of its bears.

After all, the High Arctic is a baffling and deceptive place where summer nights look like day and winter days look like night, where sometimes you cannot even tell up from down. Scientists like Derocher seek clarity, cherishing spring's eternal light, knowing that soon enough, the long, black polar night will descend, plunging them into darkness again, and come December, somewhere out there in the dark, another ice bear will be born.

Chapter 5

Ties that Bind in Greenland

East of Svalbard, along the northwestern tip of Greenland, brothers Mamarut and Gedion Kristiansen have pitched a makeshift tent on the sea ice, where the Arctic Ocean meets the North Atlantic. From their home in Qaanaaq, a village in Greenland's Thule region, the northernmost civilization on Earth, the Kristiansens traveled here, to the edge of the world, by dog sledge. It took six hours to journey the thirty-five miles across a rugged glacier to this sapphire-hued fjord, where every summer they camp on the precarious ice for weeks at a time, patiently awaiting their prey. Nearby lies the carcass of a narwhal, a reclusive unicorn-like whale with a spiraling ivory tusk. Mamarut slices off a piece of *mattak,* the whale's raw pink blubber and mottled gray skin, as a snack.

"*Peqqinnartoq,*" he says in Greenlandic. Healthy food.

Mamarut's wife, Tukummeq Peary, a descendant of Robert E. Peary, who led the first white man's expedition to the North Pole, is boiling their favorite entrée on a camp stove. They dip hunting knives into the kettle, pulling out steaming ribs of freshly killed ringed seal and devouring the hearty meat with some hot black tea.

About 850 miles from the North Pole, the people of Qaanaaq are the closest on Earth to the archetype of traditional polar life. They are the world's top predators, the human version of polar bears, and their fate—like the fate of the bears in Svalbard—illustrates how contaminants have upset the Arctic's fragile balance.

The Kristiansens live as their forefathers did thousands of years ago, relying on foods culled from the bounty of the sea and skills honed by generations. One man—a lone hunter sitting silently in a kayak, armed with a harpoon—is pitted against a one-ton whale. Such simplicity isn't quaint; it is a necessity in this hostile and isolated expanse of glacier-carved bedrock. Survival here means people live as marine mammals live, hunting like they do, wearing their skins. No factory-engineered fleece compares with the warmth of a sealskin parka or bearskin trousers. No motorboat sneaks up on a whale as well as a hand-made kayak latched together with strips of hide. No snowmobile flexes with the ice like a dog-pulled sledge crafted of driftwood. And, most importantly of all, no imported food nourishes their bodies, warms their spirit, and strengthens their hearts like the flesh they slice from the flanks of a whale or seal.

In remote villages like Qaanaaq, Greenland's hunting traditions are the strongest of all. An isolated village of about six hundred people, separated from Canada by Baffin Bay and nestled on the slope of a granite mountain, Qaanaaq faces a great sea of ice. There, the Inuit hunt seal, beluga, walrus, narwhal, even polar bear. Qaanaaq's polar night—twenty-four hours of darkness—endures from November through mid-February, and one Inuk describes it as being trapped inside a bag of coal for almost four months. Their native food helps them survive the winters, warming them from within like a fire glowing inside a lantern. When they eat anything else, instead of fire inside, they feel ice.

Greenland's Inuit eat much like a polar bear does—seal is the national favorite—and ironically, this close connection to the environment has left them as vulnerable to the by-products of modern society. Living closer to the North Pole than to any city, factory, or farm, the Kristiansens appear unscathed by any industrial-age ills but northbound winds carry to their hunting grounds the toxic remnants of faraway lands—lands they will never visit and substances they have never heard of. Their bodies contain the highest human concentrations of chemical contaminants found anywhere on Earth.

On a June day in 2001, Mamarut, groggy from sleep, opens the door of his house in Qaanaaq at 9 A.M. and contemplates embarking on a hunt. Last night, he and his brothers celebrated catching the first narwhal of the season. "What time is it?" he asks in the only language he knows, Greenlandic. "Nine o'clock? Night or day?"

Time serves little purpose here. In late spring, noon looks the same as midnight, and temperatures still dip below freezing. Yet the retreat of the polar night and return of the midnight sun means that the Arctic's treasures, long locked in the ice, are within reach again. Narwhal hunting season has begun. Every year in northern Greenland, hunters kill hundreds of the shy, almost mystical beasts. By late afternoon, Mamarut and other hunters are gathering on the edge of town to load up their sledges. As he prepares for their journey to the edge of the sea, Mamarut wears khaki pants, rubber boots, and a long-sleeved T-shirt emblazoned with "Polo Club." His wife loads the sledge with a plastic bag full of food—Ritz crackers, soup mix, bread, tea—along with some toilet paper, rope, an ax, a kayak, sealskin jackets, a rifle. The blubber of the narwhal caught two days earlier will sustain them as they camp on the ice for days. Their favorite way of eating it is raw, fresh from a kill.

A sealskin whip arches perfectly over Mamarut's dogs, as artfully executed as the line of a fly fisherman. His younger brother, Gedion, trails behind, guiding his own dog team across the glacier, dodging skyscraper-sized icebergs that rise inexplicably from the flat terrain like giant sandcastles on a beach. The ice ahead of them undulates, like waves on a frozen sea. The Kristiansens' twenty-six dogs—strong enough to pull two three-thousand-pound narwhals—race toward a sliver of brilliant blue in the distance. The waters off Qaanaaq are known to navigators as the North Water, thawed year-round in an otherwise frozen sea, where an upwelling of nutrients draws an array of marine creatures. The black and cream-colored dogs sprint across snow so bright it burns the eyes. They pant as they run, their tongues hanging so low they could almost lick the ice. Cracking the whip over their heads, Mamarut grunts at them, words that sound like *huva, huva, huva,* left, left, left. As the wind picks up, he adds two layers of clothes, including a parka crafted from the tan and gray hide of a ringed seal. His wooden sledge, a work of art perfect in its simplicity, rumbles and bumps as it carves a path in the snow, bending and flexing with the ice. The only tracks visible in the snow are from the sledges and the dogs, and the only smells emanate from their rear ends, fumes so potent they seem toxic. Part wolf, the dogs snarl at each other, eating fistfuls of raw seal meat that Mamarut and Gedion toss to them, and they are treated harshly, as slaves, not friends. This is a matter of life and death for Qaanaaq's hunters. A good dog team means life. A bad one could mean death. The hunters need senses as keen as their dogs', even keener. Spotting a black speck on the horizon, Mamarut stops his sledge. A ringed seal has poked a hole in the ice and is resting on its surface, so he picks up his rifle and blind and tromps through the snow toward it. The dogs, too eager, take off toward him, dragging the sledge. The seal hears their barking and slips back through its breathing hole under the ice. The dogs have just scared off their dinner.

Six hours after their journey began with the flick of a whip, the Kristiansens arrive at their ancestral hunting grounds. Joined within minutes by several other hunters, they gather on the edge of the ice, waiting to spot a whale's breath. "If only we could see one, we'd be happy," Mamarut whispers, lifting binoculars and eyeing the mirror-like fjord for the pale gray back of the *qilalugaq,* or narwhal. "Sometimes they arrive at a certain hour of the day and then the next day, same hour, they come back. Other times they disappear for two days." The best Inuit hunters, more than anything else, are patient. "Maybe the mother of the sea has called them back," Mamarut says with a laugh. Mamarut is big, bawdy and beefy, the elder brother and joker of the family. He celebrated his forty-second birthday on this hunting trip with some packaged chicken broth and a bit of *mattak.* Gedion is ten years younger, lanky, quiet, the expert kayaker, wearing a National

Geographic cap. The Kristiansen brothers are among the best hunters in a nation of hunters, able to sustain their families on the income from their hunts without them or their wives taking side jobs, which is unusual in Greenland. In a good year, the Kristiansens can eat their fill of *mattak* and earn more than $15,000 (U.S.) a year selling the rest to markets, and in winter, they sell sealskins to a Greenlandic company marketing them in Europe. About one-third of the food the Kristiansens consume is the meat of wild animals and birds.

Their hair is the blackest of black, thick and straight, cut short. Their skin is darkened by the sun but they have no wrinkles. One night, after setting up camp, they joke that when the sun hits their campsite, it will be as warm as Los Angeles and they will get a suntan. They are amazed to learn that their warmest days are like L.A.'s coldest, that things never freeze there. Mamarut and Gedion say they would someday like to see women on a beach wearing bikinis. Today, on a June morning, the wind-chill makes the temperature hover around zero, and their only shelter on the ice is a plastic tarp strapped to the sledge, creating a makeshift tent for five adults to sleep in, so cramped they can't bend a knee or flex an elbow without disturbing their neighbors. A noxious oil-burning lamp is their only source of heat. A camp stove, used for melting ice for tea and boiling seal meat, is set up on a wood box outside.

Along the edge of the Arctic Ocean, Gedion and Mamarut are waiting for a narwhal. They wait and wait. Then they smoke cigarettes and wait some more. Sometimes hours, sometimes days. Once they waited almost a month on the ice before catching one. During their vigil, the hunters remain alert for cracks or other signs that the ice beneath them is shifting. In an instant, it can break off and carry them to sea. No wonder that Greenlanders have several dozen word phrases for ice, but only one for tree. Ice is everything to them—it's danger, it's dinner, it's the water they drink. There's the jagged ice they encountered along the way that sliced like knives into the paws of their dogs, leaving a trail of bloody pawprints. There's the unpredictable ice that breaks into floes, sometimes taking a hunter or two along with it. There's the craggy ice along the edge where narwhals hide, in search of halibut and shrimp. A good hunter, more than anything else, knows the ice. He knows where it will shift and shatter, leaving shards like a broken dish. Mamarut jabs a sharp metal pole into the ice to check its stability. He leaves a mark in the ice, and if he sees that it has moved, he knows it is time to pack up and set up a new camp. The next morning, he announces that it is time to move onto a rocky outcropping on the ice edge. "Not safe here," Mamarut says. It is taboo in Greenland to urinate on an iceberg or kill a bear on one, but it has nothing to do with myths or evil spells. Greenland is a society heavily influenced by Christian teachings, brought there by a Lutheran minister nearly three hundred years ago. Unlike some aboriginal cultures, their beliefs about their environment are based on reality, not mythology. Like everything in Greenland, a land of pragmatics, the advice serves a real purpose: Warm liquid, whether urine or blood, will melt the ice right under your feet. Shooting an animal close enough to touch is also taboo. It's not out of fairness; it's to save ammunition.

That night, the brothers spot a pod of belugas, about ten of them, their white backs glistening in the dark water, their misty breath spraying out from their blowholes. But the belugas were there before they were, and they know there is no way to sneak up on them now. Hunters must arrive at the fjord before the whales do, then wait, like a polar bear stalking a seal at a breathing hole. Just in case, Gedion readies his sleek, eighteen-foot kayak, or *qajaq,* a wood frame stitched with sailcloth and sealskin. Although the Canadian Inuit invented kayaks, Greenlanders perfected them. Like their Danish compatriots, they are ingenious at blending form and function, in mastering the simplicity of design. Gedion inflates the skin of a seal until it looks like a seal-shaped balloon and then fastens it to a metal harpoon head with a long nylon line. When the harpoon is fired, it will float in the water like a buoy, marking the spot of a harpooned whale and preventing its head from sinking. Then he attaches a waterproof sealskin liner to the hole in the kayak that will keep him dry. A wooden harpoon, or *unaaq,* with a metal blade is laid on top. When Gedion hears or sees the whales coming, he quietly climbs into his kayak with his harpoon and sealskin buoy. He must instantaneously judge the ice conditions, the current, the wind, and the speed and direction of the whales. If a kayaker makes the slightest noise, a narwhal will hear it. Gedion knows he must first strike with a harpoon—in the whale's back, close to the fin, usually from behind so he can't be seen—then he can finish the job with a rifle. The whale must be directly in front of his kayak, about thirty feet away, close but not too close—or its powerful dive will submerge him and he will drown. Once, when Mamarut harpooned a narwhal, another one kicked him with its tail. His kayak flipped over and broke, but he managed to stay inside it and get to shore. Like most Greenlanders, he can't swim—there's not much need to master swimming when no one can survive more than a few moments in the frigid water. After a whale is harpooned, buoys are attached to its back, and its carcass is hauled back to shore, where it is butchered immediately. The big ones weigh a ton and are fifteen feet long. The hunters taste some of the blubber right away, and bury the rest in the ice. Appearing primitive, this society is, in reality, ingenious. More than a century ago, the famous explorers of the North Pole—Peary, Frederick Cook, Knud Rasmussen, Matthew Henson—learned on their expeditions through Greenland that eating Inuit food and using Inuit tools were key to survival. Peary, in particular, understood the importance of sledges, skins, and blubber as he traveled through Qaanaaq.

When asked how he catches a whale, Gedion twirls his wrist and jokes that he lassos them like the American cowboys he's seen on television. A little over a century ago, the people of Qaanaaq had no written language and had little contact with the Western world. Today, they buy dental floss and cow's milk and potato chips in their small local market, which receives shipments only a few weeks every year when the fjord has melted enough to allow ships through. They watch American films like *Nightmare on Elm Street* and *Altered States* in their living rooms on the one TV station from Radio Greenland that beams into Qaanaaq. Such amenities are increasingly popular, especially among the young people of the village. In the early 1950s, during the Cold War, their whole community was transformed when families were forcibly moved seventy miles to the north to make way for the Thule air base, an American military facility for surface-to-air missiles. In return, the government built the villagers little prefabricated wooden houses painted vivid red, blue, green, and yellow that resemble Scandinavian chalets, nothing like the igloos that foreigners assume the Inuit live in. The population of Qaanaaq, the world's northernmost municipality, has doubled since then, with Greenlanders attracted by the good hunting along the northern coast. The only civilization in the world that is farther north is a few miles away, the tiny Thule settlement of Siorapaluk, population sixty. The past few decades brought alcohol, television, and other distractions to Qaanaaq and its surrounding settlements. The changes came too suddenly, and the people were unprepared. Alcoholism, violence, domestic abuse, and suicide take a heavy toll here, having a much more immediate impact on life and death than contaminants. Often when the men of Qaanaaq aren't hunting, they are drinking. Such binge drinking is common, and it's not unusual in town to see many men stumbling around, drunk, any time of day. About 80 percent of adults smoke cigarettes, twice that of Europe, and children start early. The suicide rate is high for adolescents, and psychologists say it is because of their homeland's rapid societal transition—the impacts of modern amenities and economics—and not due to depression from polar darkness or their rugged life. "Compared with existing and known health problems in the Arctic, [contamination] is a minor issue," says Peter Bjerregaard, who heads Greenland research for Denmark's National Institute of Public Health. "There is so much violence and suicides and alcohol-related problems and so many tobacco-related diseases, which also affect unborn children." Still, Bjerregaard says, the pollution is an involuntary threat "so there is cause to be angry. Even though we can't demonstrate people dying, it is still an important issue."

While on their hunting trip, Mamarut and Gedion flipped through the pages of a Scandinavian porno magazine and read about the antics of Danish royalty. They drank British tea and smeared imported taco salsa on their seal meat. Yet the Kristiansens will never see a cow or a Christmas tree or a blade of grass. They don't know butterflies or squirrels or frogs, except in pictures from distant lands. They will never spray a pesticide or work in a factory. For anyone traveling other than by sledge, leaving Qaanaaq, or reaching it, is difficult. Until 2002, when an airfield opened there and allowed regular flights of turboprop planes, it required a costly ride on an eight-seat helicopter, scheduled once a week but highly erratic, from the U.S. government's military base to the south. Although the Kristiansens have never ventured out of their homeland to see the world, and likely never will, they know that the world travels there, riding on winter winds. Mercury, in particular, finds its way there, flowing from coal-burning power plants in Europe, North America, and Asia. Qaanaaq mothers carry the highest mercury concentrations on Earth, reaching on average 50 parts per billion in their blood, twelve times more than U.S. guidelines recommend.

The first humans arrived in Greenland around 2400 B.C., after a massive ice cap, a remnant of the last ice age, melted in Arctic Canada, freeing a path for migrants, the ancestors of today's Inuit, to head east, following herds of musk ox and reindeer. The island, wedged between Canada and Europe, was unknown to Europeans until the tenth century, when the Viking Erik the Red, banished from Iceland, discovered it and started a colony there. Today, a year-round icy shield—thicker than a mile in some places—still covers 85 percent of Greenland, the world's biggest island. Greenland is essentially nothing but rock and ice, the largest mass of land ice in the Northern Hemisphere, surpassed on Earth only by the continent of Antarctica. From above, jagged shards of ice float in the blue ocean, as far as the eye can see. The sea looks dark, hard, almost solid, as if nothing lies below its surface, belying its riches. Etched by glaciers, Greenland is ever-changing. Ice sheets melt and freeze countless times, scouring the granite bedrock and slicing deep fjords into the landscape like sheer cuts of a knife. The land is rising, emerging from the sea at the rate of a few feet every century. The polar bears and other animals that live here have adapted to this harsh environment, and over the centuries, people have adapted, too. Mercury has always been a natural element in the Earth's crust, which means that the bodies of humans and animals have always contained traces. In 1972, the well-preserved, mummified remains of two Thule children and six women who died in the 1400s were found in a gravesite at Qilakitsoq, an old settlement in northern Greenland. Their bodies contained traces of mercury, and an analysis showed they ate seal, caribou, hares, and auks—the same foods that Greenlanders eat today. Yet the mercury concentrations found in Greenlanders are much higher today as human activities have caused mercury emissions to surge.

Greenland is dominated by the sea—its food, its weather, its economy. To Americans, the sea is more recreation than sustenance. But Greenland is a land where no crop grows, no livestock survives. Its people are dependent, more than anywhere else on Earth, on the sea's riches. Greenland has no trees, no fertile soil, virtually no grass, which means no cows, no pigs, no chickens, no grains, no vegetables, no fruit orchards. In fact, there is little need for the word "green" in Greenland. Instead, the ocean is the food basket for Greenland's 57,000 people. In the remote villages such as Qaanaaq, people dine on marine mammals and seabirds thirty-six times per month on average, consuming about a pound of seal and whale each week. In larger towns, such as Nuuk and Disko Bay, people eat about half as much because there are more imported food options. Seal is the Inuit's favorite meal, with 160,000 seals eaten in Greenland each year. Everything else, from tea to bread to cheese, is imported from Denmark. Imported food is expensive, often stale, and not very tasty or nutritious. In Nuuk, where the average family income is relatively high, about $23,000 (U.S.), a box of cornflakes in 2001 cost $4, a carton of eggs $3, a pound of hamburger $5, a four-pound chunk of lamb $35. In Qaanaaq, family incomes are only half as high, averaging under $13,000 (U.S.), yet food is just as costly. In Canada's province of Nunavut, across Baffin Bay from Greenland, store-bought food for a family of four would cost $240 per week, or $12,500 (U.S.) per year—an amount unaffordable for nearly everyone there, as the average annual family income is under $35,000, according to a 2003 report by Canada's Northern Contaminants Program.

Jonathan Motzfeldt, who was Greenland's premier for almost thirty years and is now vice premier, decorates his office with functional and stylish Danish furniture. Nuuk, the town in which he lives, resembles a miniature suburb of Copenhagen. He wears Western clothes and is influential in Danish politics. Yet on the wall of his office, he has hung four wooden harpoons. They are not simply artifacts, but a testament to his country's hunting culture. Hunting isn't sport for his people; it's survival. About 85 percent of Greenland's population, 48,000 people, are Inuit. "We eat seal meat as you eat cow in your country," he says. "It's important for Greenlanders to have meat on the table. It is as essential for our culture as Kentucky Fried Chicken is in America."

In an adjacent government office, an advertisement for Chick-Fil-A, a U.S. fast-food chain, hangs on a wall. It's a photo of three cows holding a sign saying "Eat More Chicken." If Greenlanders were holding the sign, it would say "Eat More Whale." They cannot comprehend the American diet. They don't understand why Americans eat Whoppers and KFC and bacon cheeseburgers. Whether it's chicken, cows, or pigs, they view it as unhealthy food raised in an oppressive environment, full of bacteria and contributing to an unwholesome lifestyle. It's their greatest irony to discover that their food has been tainted by what they perceive as the excesses and waste and greed of the industrial world. Motzfeldt worked on a pig farm in Denmark one season and, horrified by what he saw, he vowed to never eat pig meat again. He prefers the animals he eats to live wild and free, as nature intended, and die while waging a fair fight with a hunter. Motzfeldt is proud of his homeland's blend of the old and the new, and he is confident the traditions will survive. "Nothing is divine here in life but we as Inuit and Greenlanders have survived for many, many generations on the resources harvested from the richness of nature, whether it is fish or seals or whales or birds or polar bears," he says. "And as a culture where you harvest the surplus of nature, people in other parts of the world have to acknowledge this is a sound way of life. We do conduct scientific research on our environment's well-being to be sure we do not overexploit specific resources, and we integrate imported foods. But hunting seals and whales is essential for us to survive as a people." The Inuit, according to the International Whaling Commission, are "the most hunting-oriented of all humans."

Lars Rasmussen, a fifty-two-year-old hunter in Nuuk, Greenland's largest town, has stopped at the town's cultural center to discuss the importance of hunting. He comes from generations of hunters—his father, his grandfather, his great-grandfather, as far back as he can remember. Hunting provides more than half his income, while fishing provides the rest. He considers himself a hunter, not a fisherman, but he knows he has to fish for Arctic char and salmon to provide enough food for his family. "Eighty percent of what we put in the pot comes from my own catch," he says. "We are living in a place that is very cold and it's not by accident we eat what we do. We are not able to survive on other food."

It is early June, about 40 degrees Fahrenheit, and the sun is shining, day and night. Four months of darkness, four months of light, four months of both—this is the cycle that guides Rasmussen's life. He wears a gray wool sweater, tennis shoes, and the blue bibs of a hunter, with a baseball cap over jet-black hair. He has the hands of a fisherman and the warm, round face of the Inuit, framed in eyeglasses. Soft-spoken, he speaks only Greenlandic. "I have a message," he says softly. "I want people out there to be aware that hunting is so important to us, so fundamental, that we will not be able to survive without it. Outsiders who do not live here have so much influence on our daily lives with all the restrictions they press our government to put on us. That hurts our livelihood and culture. We don't think it's right that other people have so much influence on our lives. That is my message."

He has heard about the contamination of his people on the news—the local radio station or newspaper reports something about it maybe twice a year. But it is quickly forgotten. In dietary surveys conducted in Canada, most Inuit say they have not altered their diet to limit exposure to contaminants. Greenland's home rule government has issued no advisories and doctors continue to tell people—even pregnant women—to keep eating their traditional food and nursing their babies with no restrictions. "I'm not concerned," Rasmussen says. "But I suppose I and my children in the years to come will learn more."

"I feel more powerless than anything," he adds. "It's so big an issue and the polluters are coming from all over the world and our sea here is the ultimate dump. I don't see how we can do anything about it. Would they listen? They have big money. We feel very much powerlessness."

Outside the wind is howling. The sea is choppy with whitecaps, too dangerous for hunting boats to head out today. Rasmussen grabs his cell phone, zips up his parka, and returns home.

In his town of Nuuk, population 13,500, just south of the Arctic Circle, people don't need whale meat for subsistence, unlike Greenland's remote villages. If Nuuk is all you see of Greenland, you would wonder why traditional foods are so important. The town looks nothing like Qaanaaq or the North Pole. Nuuk residents can walk into the cultural center and order a Coke and a shrimp salad sandwich for $8. They wear Polartec fleece, not polar bear fur. They watch American-made blockbuster films such as *Lord of the Rings* or *Pearl Harbor* only a few days after their U.S. debuts. Schoolchildren ride razor scooters and carry cell phones. But even in Nuuk, there are few places on Earth where hunting is so vital to a culture.

Ingmar Egede, an educator fluent in three languages, founded an international center to train indigenous peoples around the world to participate in global affairs but his life in Nuuk remains steeped in Inuit tradition. His home sits on the edge of a fjord brimming with sea life, and on late summer nights, Ingmar slumbers to the sound of whales exhaling as they migrate past his window. "You can go out and pick your dinner at any time," he says. Whale meat has always been a tonic for Ingmar. Upon returning from his worldwide travels, he grabs a piece from the freezer and his stomach calms within an hour.

Although he is not a hunter by trade, Ingmar took part in many hunts when he was younger, landing "hundreds of seals, many walrus and belugas." When he hunted, he always spoke and thought in Greenlandic, the Inuit language, not in Danish or English. The feelings just didn't translate into a modern language. The hunt unleashes a lot of primitive emotion. The chase is thrilling, and getting close enough to a one-ton wild creature to fire a harpoon makes your heart pound with adrenaline and aggression. The environment is cold and harsh, the payoff is uncertain. There are few words in English to express the joy of the hunt. And there are few Americans who don't cringe at the thought of it. To Americans, a chicken is only a chicken. But a whale is . . . well, godly, sacred, too noble for humans to eat. To the Inuit, that is absurd. You should respect what you eat, Ingmar says. Hunting, he says, is more ethical than raising livestock. How can you treat an animal so badly, putting it in pens, as if its only purpose in life is to be jailed and slaughtered, and then have respect for yourself? Whales and seals live wild and free, and are given a fighting chance against a lone hunter. "It's a lack of respect for yourself to eat animals that are so mistreated," Ingmar says. "Why? Because you are what you eat." He tells the story of an American surgeon who couldn't kill a wounded rabbit hit by his car. He spent his life slicing open people but could not wring a suffering rabbit's neck. Is this a culture that respects itself, Ingmar asks, or just one that respects animals in a pretense of respecting itself? "Americans are so estranged from where their food comes from. Their food is in boxes," he says. Chickens suffer "ten million times a day" compared with fewer than two hundred whales that are killed every year in Greenland, Ingmar says. He contemplates all those chickens and throws up his hands. 'How do you measure suffering?" he says. "A chicken suffers as much as a swan."

The differences in attitudes between Americans and Greenlanders are so profound that they are irreconcilable. Americans come from a catch-and-release society and sport I BRAKE FOR ANIMALS and SAVE THE WHALES bumper stickers on their cars (although each American, on average, eats more than 200 pounds of cow and chicken meat a year). Greenlanders hunt whatever they can and eat whatever they kill, treating no animal with any more respect than another. The two cultures tolerate each other in international forums, setting whaling quotas, goals of sustainability, and standards of humane treatment. But they will never really see eye to eye.

Every year, Greenlanders are allowed to kill up to 187 minkes and 19 fin whales under the rules of the International Whaling Commission (IWC), which also dictates uses of powerful grenade-equipped harpoon guns and high-powered rifles designed to kill quickly. Neither species is endangered, and minkes number in the hundreds of thousands in the North Atlantic. Major environmental groups, even protest-prone Greenpeace, no longer oppose whale hunts by the Inuit as long as they are well managed under international law.

It is ironic, Ingmar says, that the Arctic's animals are being contaminated by the very people who say hunting whales is wrong. The pesticides and industrial chemicals drift from industrialized nations that try to stop the hunting of whales but do little to stop their own poisoning of them. Anger, he says, is the overriding emotion among Arctic people who are aware of the pollution. Ingmar, unlike most of his fellow Greenlanders, travels a lot, so he has heard the scientists' warnings about hermaphroditic polar bears and damage to children's brains and immune systems.

"The chemical threat is the ultimate threat to mankind—worse than bombs and war," Ingmar says. "You cannot hide from it. It reaches everywhere in the world."

A thick fog has moved into the harbor off Nuuk, shrouding the low-slung, snowcapped mountains. Tones of gray are everywhere on a foggy day like this. An icy rain pelts Ujuunnguaq Heinrich as he stands at the bow of his boat. Ujuunnguaq rests his gloved hand on his harpoon gun as he scans the horizon for signs of a minke whale. One crew member hangs atop the mast of the old, peeling, thirty-foot wooden boat, on the lookout for ripples in the sea that indicate they have found their prey.

An hour into the voyage, Heinrich spots not a whale but a seal, its head bobbing on the surface. He aims his rifle, the same high-powered weapon used to kill African elephants, and pulls the trigger. A deafening blast shatters the foggy silence and reverberates across the water. He misses, then fires again. The seal dives beneath the surface and escapes. It is a jarring reminder that this is no mere fishing voyage, no whale-watching trip. Hunting marine mammals is a remnant of an ancient way of life that is now rare outside the Arctic. One of 120 full-time hunters in Nuuk, Ujuunnguaq, thirty-eight years old, has killed twenty minke whales over the past ten years. He fishes for halibut and kills about two hundred seals per year, but whale is the trophy, the ultimate prize, its two tons of meat worth $10,000 (U.S.), which he splits with two crew members. He shares the whale meat with his family and friends, and his family eats it several times a week during whale season. He sets some aside for his wife to fix for special holiday meals. The rest is sold to other Greenlanders, who are prohibited by the IWC from selling it to other countries.

The spring rain freezes, turning to snow. Four hours after they set sail from the harbor, Ujuunnguaq declares the hunt over for the day. He and his crew return to their families, empty-handed. "This is my life and I don't want it different," Ujuunhguaq says in Greenlandic. "To chase one, to hit one, to bring one home. That is the essence of life for me."

Under international restrictions, Greenland's hunters aren't able to provide enough whale meat and blubber to meet their own people's dietary needs. Traditionally, for centuries, they hunted much larger whales—humpbacks and bowheads, which were virtually wiped out by European commercial whalers in the eighteenth and nineteenth centuries. Today they are left only with minke whale, fin whale, and the smaller narwhal and beluga. Amalie Jessen, Greenland's representative on the IWC, says hunters would need at least 670 tons of whale meat to meet the needs of people in West Greenland alone. But the take throughout Greenland is no more than 558 tons, so there is a shortfall of meat. Fin whales replaced humpbacks for hunters, but the seventy-foot whales are too fast and the boats too small to catch them. For minkes, however, Greenland is always seeking to increase its quota and obtain permission to sell its meat to other nations, particularly Japan, in an effort to gain economic self-sufficiency. Independence is of utmost importance to Greenlanders. Although part of Denmark, they won home rule in 1978. Even their cultural center isn't called a Nordic House as the centers are in the rest of Scandinavia. The Danes, Swedes, Norwegians, and Finns had to tread a fine line during its design and construction to ensure that it was embraced as Greenland's own.

Hunters in Nuuk have learned about the PCBs and other contaminants from Danish media; they are angered by the intrusion but do not worry much about it. "People say whale and seal are polluted, but they are still healthy foods to us," Ujuunnguaq says. "We hear about the contamination and it's only words for us because we've been eating it for generations and don't see any impacts." Why should they have to change their traditional ways because the rest of the world is spreading its poisons? Just as poisonous, perhaps more so, they say, are the Western attitudes toward whale and seal hunting.

Ujuunnguaq has more pressing concerns than chemicals. He wishes he were allowed to kill more whales. He wishes he didn't have to equip his harpoon gun with grenades, which are expensive and waste valuable meat. He is fully aware of the Western attitudes toward whaling, and he lives every day with their consequences. As a hunter, he knows he has to weather the whims of nature but he also has to weather the whims of society. "Why should outsiders get involved in our livelihood?" he asks. "Why should they tell us what we should eat?" He jokes about the humpback being America's poster child. He respects whales as powerful adversaries and magnificent creatures but he doesn't understand the sentiment.

"We don't have cows. We don't have pigs. Whales are what we eat," he says. "Americans give whales nearly a holy role, but it's written 'Thou shalt only have one God.' And it's not a whale."

Mamarut and Gedion Kristiansen, like most of the people of Qaanaaq, don't travel outside the fjords surrounding their village, and they remain oblivious to the scientists and political leaders fretting about how many parts per billion of toxic chemicals are in their bodies. They simply don't have the luxury to worry about threats so imperceptible, so intangible. Instead, they worry about things they can hear and see: thinning ice conditions, the whereabouts of whales, where their next meal will come from. Anxiety about chemicals is left to those who live in faraway lands, those whose bodies contain far less of the substances. The Kristiansens learned a little about the contaminants—the *akuutissat minguttitsisut*—from listening to the radio. But they have not changed their diet, and no one has advised them to. Seal, narwhal, and beluga are what they hunt, so it is what they eat, despite the high contamination. Virtually every day, they eat the meat and *mattak,* and with every bite, traces of mercury, PCBs, and other chemicals amass in their bodies. "We can't avoid them. It's our food," Gedion says with a shrug.

On this five-day hunting trip, a short one for the Kristiansens, they reap little reward for their patience—a few seals, two auks, and two eider ducks. Mamarut, his wife Tukummeq, and Gedion pack up their sledges and drive the dogs back toward Qaanaaq. "Sometimes you have to just go back empty-handed and feed your dogs," Mamarut says.

Upon returning to their village, Inuit hunters share their experiences so that everyone may learn from them. The Kristiansen brothers learned to hunt narwhal from their father. Now Gedion's son, Rasmus, four at the time of this hunt, often joins their hunts, amusing his father by pretending to drive the dogs and harpoon a narwhal. It won't be long before Rasmus will paddle a kayak beside his father. Since around 2400 B.C., this Inuit legacy has been passed on to generations of boys by generations of men.

Their ancestors' memories, as vivid as a dream, as old as the sea ice, mingle with their own, inseparable.

"Qaatuppunga piniartarlunga," Mamarut says.

"As far back as I can remember, I hunted."

Chapter 6

A Fish Can't Feed a Village: Alaska's Communal Hunts

On the other side of the Arctic, along the North Slope of Alaska, the art of whaling has been handed down for five generations in Susan Patkotak's family. Her grandfather about a century ago taught her father to hunt the giant, fifty-ton bowhead, then her father taught her husband, and he, in turn, taught their sons and grandsons.

At eighty-one, Susan is arthritic and bedridden at a nursing home in Barrow, and her husband, Simeon, sixty-eight, has turned over the harpoon to his sons although, as patriarch, he remains captain of his family's crew. Nevertheless, the spring hunting season is no less important to the elderly Patkotaks. As the wife of a whaling captain, Susan played a critical role in the hunt for many years. While the men camped at the edge of the ice, sometimes for days or weeks at a time, she would supervise the women's activities, ensuring that the Patkotak crew had all the food and supplies and support they needed. To Susan, who has twelve children, thirty-four grandchildren, and fourteen great-grandchildren, whaling means pride in family, service to the community. The captain is hailed as a hero, his generosity, courage, and leadership unmatched by anyone in town. Here in the northernmost point in the United States there is no more joyous occasion than a successful spring hunt. No wedding, no birth of a child, no anniversary can compare to the rejoicing that follows the landing of a bowhead. In the old days, when a crew caught a whale, the captain would dispatch a runner carrying the family flag. He would run all the way from the camp to town, miles and miles, waving the flag and broadcasting the news to the whole community. The tradition remains. The only difference is that the runners return to town on snowmobiles rather than on foot.

On a morning in May, a young runner, still in his white hunting parka, knocked on Susan's door at the nursing home. Her youngest son, Crawford, had landed the first whale of the season, and as the matriarch of the family, Susan would be the first one to know. The runner arrived carrying the Patkotak crew's flag, the same design her grandfather's crew hoisted seven decades earlier. The news meant that everyone for miles around would have ample meat and blubber through the long, dark winter, when temperatures plunge to 60 below zero.

In Qaanaaq, Greenland, whaling is a personal triumph—one man, one narwhal—but in Barrow, because of the immense size of the prey, whaling is a communal event. Along Alaska's North Slope, a fish feeds a family but a whale feeds an entire village.

"Praise the Lord!" Susan cried out. "God created the whale. Praise God!" And she started to laugh and cry.

Preparations for the feast, the *nalukataq*, were about to begin.

It is June 22, 2001, the official start of summer and the day when Eugene and Carl Brower, cousins from the most famous of all Barrow families, are hosting their *nalukataq*. On a dirt field on a beach overlooking the Arctic Ocean, a windbreak had been set up early that morning. Pickup trucks arrive, hauling dozens of cardboard boxes. Each is filled with whale meat and *maktak* that had been stored in the Browers' ice cellars.

When you invite a thousand people to share a meal that lasts all day and all night, the preparations take many weeks. The members of Eugene Brower's *nalukataq* team, supervised by his wife, Charlotte, cut up the meat and *maktak* and box it up. They pluck and skin the geese and ducks for soup, cook the berries, bake rolls and cakes. They build the windbreak and the wooden posts that will hold the skin for the blanket toss. The crew has already retrieved its share from the whale's belly, stripping off a belt of meat around its waist, but most of the rest goes to the community. A captain hosts not just one feast, but four. Hundreds of people already traipsed through Eugene and Charlotte Brower's home on the day after the catch to sample the meat and blubber, and they will return for Thanksgiving and Christmas, too. Every year, they wear out Charlotte's linoleum. Catching a whale is expensive; the feasts probably cost each captain $20,000, but the family considers it an honor and a duty. "What you get in return is nothing monetary," Charlotte says. "You have a feeling inside that you've done something for the community."

About six dozen boxes of meat and blubber are unloaded by Eugene's team, all of it from a single whale. Then Carl's crew arrives, unloading more. The "Taalak" crew, headed by Carl, and the "Aalaak" crew, headed by Eugene, each landed a whale this season. Both Eugene and Carl are descendants of Charles Brower, a Yankee whaler who settled in Barrow in the 1880s, learned the language, adopted traditional hunting skills, opened a trading post, married an Inupiat woman, and spawned what has become one of the largest and most renowned Eskimo families in Alaska. Photos of whaling crews at Point Barrow back in the 1890s look hauntingly familiar, as if they could have been shot today. Today's hunters have cell phones and snowmobiles and polar fleece but they also still have harpoons and skin boats and polar-bear clothing and sealskin floats. They know, as Charles Brower came to realize 125 years ago, that nothing could surpass the know-how of the North Slope's native hunters. So much, yet so little, has changed in Barrow.

Every spring, as the sea ice retreats, bowhead whales appear off the coast of Alaska's North Slope on the way to their summertime feeding grounds, and the Inupiat hunters are ready for them. They have scouted the melting ocean, prepared their sealskin boats, guns, and harpoons, blazed a trail through the ice, and set up precarious camps on the edge of the sea. A hunting crew can't take a bowhead alone. Just hauling the whale to shore takes a village, and butchering it takes days. In June, right at the summer solstice, the entire village gathers on the beach to share the meat and *maktak*. A bowhead is immense—its mouth alone can be three times the size of a man—and a single carcass can provide a mountain of meat—as much as fifty tons of food that sustains thousands of people throughout much of the year. The Inupiat (most still call themselves Eskimos) are Americans, so they inhabit a nation where many people don't even know their neighbors' names, much less feel any desire to feed them. But in Barrow, no one goes hungry. No money ever exchanges hands. No questions are asked. No stranger is turned away. People eat until they are full, and then they fill their ice chests with meat and *maktak* to take home. Sharing is the heart and soul of a resource-dependent society like Barrow's.

A century ago, the Inupiat shared because there were no jobs and food was scarce. Today, the meat is less vital, but grocery store food is so costly here that most people living along the North Slope still need the meat to survive. Three out of every four residents are subsistence hunters. No roads lead to Barrow. Ships carrying provisions can reach here only a few weeks out of the year, when the ocean melts. The land underneath the town is permanently frozen, and the North Pole, at 90° north latitude, is a mere 1,129 miles away. Bowhead is not just a delicacy here; it is the people's major form of protein, the mainstay of their diet. There are other foods to eat in Barrow: beef flown in from Anchorage, salmon sent from the distant shore of southern Alaska. But the bowhead whale is more than sustenance for the Inupiat. It is the embodiment of their history as well as their future, the lifeblood of their grandparents and their grandchildren. Like the Browers, many of the 4,500 people of Barrow have the comingled blood of nineteenth-century Yankee whalers and the native Inupiat. The North Slope's oil economy—crude was discovered in Prudhoe Bay, to the east of Barrow, in 1968—has left them fairly well off compared with most other Arctic people. Barrow residents drive pickup trucks and SUVs, and vacation in Anchorage or Seattle or even Honolulu, and many of their offspring leave this unattractive outpost to head off to southeastern

Alaska or the lower forty-eight for college or jobs. Muddy in spring, dusty in summer, Barrow is far from alluring. The sky is often a drab gray. Litter abounds. Rusting cars are parked in dirt yards. Houses are weathered and peeling. The land is treeless, barren, the dogs scraggly. The place itself seems to have no soul. Those who visit Barrow may wonder why anyone would stay in such a forsaken place. But it's the heritage of hunting—the gathering and preparing of traditional foods—that brings this town to life. Few American towns are imbued with such a communal spirit. Every spring, when the Arctic Ocean begins to melt, the Inupiat of Alaska are enticed back to the sea, just like their more archetypal polar compatriots in Greenland. For a month, sometimes two, they watch and wait on the edge of the ice near Point Barrow, searching for ripples in the water that tell them the bowhead has returned. Only then can they fulfill their destiny.

The Inupiat never boast that they have caught a bowhead. They *receive* a whale, and they do so humbly. They believe the whale chooses to sacrifice itself, that if a hunter creates a place pleasing to the animals, it will choose to die there and tell others to come next year. Nothing logical about animal behavior or natural selection can shatter this belief, it is so deeply felt among the Inupiat. Doubting it would be akin to blasphemy, an abandonment of their heritage. A month before the *nalukataq*, Eugene Brower—a whaling captain for almost twenty years—watched a large bowhead shove a smaller one, a thirty-five-footer, to his skinboat, offering it to him and his crew. Its remains are now boxed up, ready to be feasted upon. Respect for the animals, he says, is paramount in their hunts. He remembers a dream of his father's more than a decade earlier: Harry Brower Sr. dreamed of a mother whale and baby and a ropelike object. It turned out that, unbeknownst to him, one of his sons had just caught a mother bowhead—there was a calf inside—and when a crew hauled it ashore, a pulley broke and hit his son in the stomach, injuring him. It was considered a message, a warning to hunters. Since then, no one in Barrow ever hunts beyond May to avoid capturing pregnant whales.

"The bowhead," says Eugene Brower, who presides over Barrow's whaling captains association, "is the greatest animal that God has ever made." It is obvious that God expects us to live off the whales and other animals of the sea, he says. Why? Because there is nothing else here, he says, spreading his arms toward the icy shores of the Arctic Ocean. Kenneth Toovak Sr., born in 1923 and a lifelong resident of Barrow, joined his father's whaling crew as a boy, back when crews used dog teams instead of snowmobiles. God, he says, has provided for the Inupiat. "After the Lord made Earth and heaven and the moon and stars, then the animals were made by him on the Sabbath day. He looked at what he created and said 'it looks good' and now said 'you take care of what I made for you.' They were made available for us because up here in North Slope, the sun doesn't shine at all times. When it becomes fifty below and the wind blows cold, that's the time a person needs red food." He remembers in the 1960s when the federal government tried to send villagers beef to replace bowhead. Each family was sent about ten pounds, enough for a meal or two for a large extended family. Toovak remembers thinking, "What good does that do us?" A single whale can provide 60,000 pounds or more. Imported meat and other foods are available, but they cost perhaps 50 percent more in Barrow than on the mainland. Apples were $6 per pound in 2001, a gallon of milk $7, a pound of butter $5.50, and a pack of bacon almost $8. "Subsistence hunting is important here because you'd be continuously broke if you tried to live off the grocery store," says Dennis Packer, chief administrative officer of the borough.

At the beach, by early morning, the skin of a bearded seal removed from one of the captain's boats has been strung on the wooden posts, creating an Eskimo trampoline. One by one, teenagers climb onto it as their friends ring its perimeter, stretching the skin to toss the jumper and challenging each other to leap higher and higher. Someday, some of these boys will supervise their own whaling crews, hosting their own blanket tosses. The flags for the two Brower crews are flying at the fence—Carl's a yellow X on a bright blue field, Eugene's a white diamond on a background of red and green checks. The two families sing gospel songs, then form a circle around the tables, holding hands for a prayer, a mix of Christian and Inupiat rituals. Carl's mother and Eugene's aunt, Jane, steps up to the microphone, says the work is hard but she is thankful for the hunters and the animals, and they are distributing the food to everyone today to spread their happiness. About three hundred people have gathered for the noontime meal of goose soup, a rich, aromatic broth of meat and rice. The family members ladle it out to each person from big metal pots, passing out bread and rolls, too. Then, the innards of the whale are served—intestines, kidney, tongue, heart. Wasting any part of a bowhead is considered disrespectful to the spirit of the animal they slaughtered. With twenty bowhead caught, this was the best season in nearly half a century, and Barrow will celebrate with seven feasts. All seven days have been declared borough holidays, with schools and offices closed. "The only true holiday is when a captain receives a whale through his boat and lands it," Eugene says.

By 5 P.M., several hundred people, from seniors in wheelchairs to babies bundled in blankets, have assembled at the beachfront feast, toting their ice chests. It is the coldest summer feast in recent memory, temperatures below freezing and windy and overcast, and people are huddled in parkas next to the windbreak. Numbers are posted over their seats. One Brower crew will serve the even numbers, one will serve the odd numbers. The adults break up the kids, shooing them away, and pull down the blanket toss. They are preparing for the best part of the celebration—the meat and *maktak*. It has been stored in the month since the hunt in an underground ice cellar—a hole dug deep in the ground, through the permafrost. It comes out fresh, with no freezer burn whatsoever.

"Good evening," Jane says into a microphone. The two crews and their families—about fifty people—hold hands in a circle around the metal tables piled high with boxes of meat as an elder prays, offering a blessing to thank God for the successful hunt. The crowd cheers, then the families fan out with boxes and tubs. In the first round, each guest gets a fist-sized hunk of *maktak,* pale pink blubber with black flesh. Most eat it right away, dicing it into small cubes, sprinkling it with salt and savoring it. Young girls come by, pouring hot tea and coffee from pots. Some have brought bags of potato chips and Cheetos, but it's the *maktak* they crave.

"If you didn't get any *maktak,* raise your hands," a Brower announces into the microphone.

Then another round comes, dropped into plastic Ziploc bags that are stashed into each guest's ice chest. Then a third.

"Enough already," says one guest. "No, take more," a crew member says.

"Who didn't get thirds?" someone announces. "Anyone else? Speak now or forever hold your peace." Some guests have so much meat it overflows their ice chests. Some store it for the winter; some barter with friends and relatives elsewhere, exchanging it for salmon or caribou.

It takes a full hour, until 7 P.M., to distribute the three rounds of *maktak.* Then comes the meat, one round, then another, handed out from big blue tubs the size of laundry baskets. Then comes another round, this time, four long strips of flipper per family. Then another. Finally, the boxes are empty. Then comes the dessert—two-tiered strawberry and chocolate cakes served with sweet pink ice cream made from caribou fat and pieces of meat. It is time for the adults to start their celebration. The blanket toss is set up again, this time suspended eight feet above the ground. The crowd gathers around, and one by one, the crew members climb aboard, jumping on the skin. The athletic ones perform aerial acrobatics, leaping ten feet up, high above the crowd, tossing handfuls of colorful candy that are scooped up by the children. Even women and elders take a turn. Everyone gets a chance, and it lasts for hours. Usually, sometime during the feasts, an ambulance arrives, picking up someone who has made a bad landing and injured an ankle or a knee. This night is no exception; an elderly hunter has wrenched his knee. On most *nalukataq* nights, everyone retreats to the school for Inupiat dancing—more like moving story-telling than dances—led by the captain and his crew. But this particu-lar night, because there was a funeral in town, there will be no dancing. In summer, the sun never sets, so the adults, even some children, lin-ger at the blanket toss past midnight.

The Brower cousins fed more than six hundred people at their daylong feast, a smaller turnout than usual because of the cold, windy weather. This summer, there is enough whale to fill everyone's freezer in Barrow, and then some. Their spring quota, set by the Interna-tional Whaling Commission, is twenty-two whales. The Inupiat have hunted bowhead for thousands of years but it wasn't until the mid-nineteenth century, when commercial whalers moved in, killing an es-timated 18,000, that the whales began to vanish, coming close to extinction. Bowhead were hunted by foreigners and Yankees not for food but for their baleen, a tough, elastic material in their jaws, called whalebone, that was coveted for corset stays, umbrellas, fishing rods, and a variety of other purposes—until someone invented flexible steel and the whalebone trade collapsed by 1908. In 1921, the commercial whaling ships took their last bowhead. By then, the species was in trouble, and they had wiped out the real giants, the seventy- to eighty-footers. Now the biggest bowhead in Alaska measure only about fifty-five feet long, and the species has not yet recovered, remaining on the U.S. endangered species list. Nevertheless, American environmental groups recognize the cultural importance of Inupiat whaling and do not oppose it. Since 1972, the Sierra Club has favored the prohibition of all hunting of marine mammals, with the exception of hunting by native groups. Still, the Inupiat hunts are internationally managed, with the hunters at the mercy of the International Whaling Commission, which sets the aboriginal quotas for all large whales.

In 1977, the IWC banned the Alaskan hunts after scientists, with-out consulting the Inupiat, estimated that only six hundred bowhead existed after trying to count them. The hunters—but not the scientists—knew that bowhead hid underneath ice floes and smashed through the ice to breathe. The population estimate was ludicrously low, but at the time, the Inupiat had little political clout so the ban prevailed for a season, until the scientists learned they were off by at least a factor of ten. It was a dark time in Barrow, socially and economically devastating. Marie Carroll, a health worker in Barrow, says it was the first time she ever saw grown men cry. To this day, no one has forgotten the ban of 1977, and bitter feelings remain. Censuses are now conducted every four years using under-ice acoustics, and an estimated 9,000 bowhead swim the Chukchi, Beaufort, and Bering seas, their numbers increasing by 3 percent yearly. In the communities along Alaska's North Slope and in Chukotka, Russia, across the Bering Strait, the IWC allows no more than sixty-seven bowhead to be taken each year. Occasionally still, some nations, trying to get political leverage for their own causes, try to suspend the hunts.

The industrialized world has left its mark here in another way, too, and Eugene Brower worries about the spread of toxic pollutants. The whales they eat contain an array of chlorinated chemicals, including PCBs, toxaphene, DDT, chlordane, and other pesticides. But because bowhead eat zooplankton and are low on the food web, their concentrations of organochlorines are about one-tenth as high as the toothed narwhals and belugas eaten by the Inuit of Greenland and Canada. Also, because of prevailing winds and currents favoring transport to the eastern Arctic, the chemicals tend to accumulate in Greenland, Svalbard and eastern Canada rather than in Alaska. With relatively low concentrations found in their prey, Alaskan scientists led by Todd O'Hara of the North Slope's Department of Wildlife Management suspect that there are no adverse effects on the Inupiat. But they warn that the contaminants fluctuate seasonally, as the bowhead migrate between the Bering, Chukchi, and Beaufort seas, and that levels of one pesticide, chlordane, are high enough to justify monitoring how much people eat. The Alaskan government has issued no food advisories in the North Slope, and while the Inupiat are aware of the threat, they worry more about the bowheads' health than their own. They have noticed more tumors and diseases in the animals they hunt, and some populations are thinning, although no one knows whether contaminants are causing the problems or perhaps climate change or some natural stress. One scientific team linked the chemicals to immune suppression in Alaska's fur seals, but there is little other evidence of effects there, in animals or people. Still, the Inupiat are wary of anything that threatens their marine environment. "This is our garden," Eugene says, "and we must protect it."

Charlie Hopson and his crew have spent two weeks in subzero temperatures, hacking through the ice with hand picks, building a trail five miles long that snakes from town to the edge of the ice, where the bowhead feed as they migrate between the Bering and Beaufort seas. This is Charlie's fiftieth season whaling. He started at a younger age than most, running errands for his grandfather's crew when he was nine, and since then he has been involved with the catching of well over fifty whales. For the past fifteen years, he has been a whaling captain, one of Barrow's most successful.

On a Saturday morning in May, as the spring hunt of 2002 draws to a close, Charlie climbs onto his snowmobile and hits the trail. He zigzags across the ice, dodging giant, blue-hued chunks sprinkled on the surface, reflecting the sun's rays like giant prisms. It takes about an hour to travel the five miles to his camp.

"This is my spot," he says. He calls the frozen icy shore the "white sands of the Arctic." From here, the icebergs offshore look strangely like a city skyline. This is the place Charlie has chosen to hunt for whales as they round Point Barrow on their northbound spring migration. He eyes the currents, checks the winds, smokes a cigarette. Then he climbs back on his snowmobile and heads back home to call his crew, to wake them up, to tell them he's ready.

"Now," he says into the telephone.

He's worried about the winds. He can tell from the flocks of eider ducks—hundreds strong—sailing over his head that they're about to change directions. Once the winds switch, they'll blow dangerous icebergs toward his camp. He predicts he has only twenty-four or thirty-six hours. Charlie packs up a thermos of tea and chews on a piece of *maktak*. The radio crackles. The wives and mothers are saying good morning to their crews out on the ice in a mix of English and Inupiat. Someone prays that the crews will be able to go out today.

It's a stressful time for a whaling captain. The captain's hardest job is looking out for the safety of the whole group. He has a bare-bones crew of five this season. Emma, Charlie's daughter, usually hunts with him, but she is pregnant now with her first child. He expects her to become a whaling captain some day, the first female one in Inupiat history. His son, a mechanical engineer at the lucrative oil fields in Prudhoe Bay, usually joins his crew, too, but he is preparing to return to work soon.

Bacon and French toast are sizzling in a pan, and his wife serves them to him before he heads out. He slides a rifle into a slot on his snowmobile. By 11 A.M., he is back on the trail. Snowmobiles replaced sledge dogs here long ago, in the 1960s. He stops midway on the trail, where his umiak, his wooden frame boat covered with the skins of seven bearded seals, is waiting. It is brand-new, and still reeks of fresh skins. Even at twenty-five feet long, one of the biggest in town, the skinboat is only half the size of some bowhead. He is the first to arrive, so he sits on a wood sledge to wait for his harpooner, his brother William. It's about 10 degrees Fahrenheit; he dons a parka, its hood and sleeves rimmed with brown wolf fur, and covers it with a white shell so he blends in with the ice and won't scare off the whales. He wears a Seattle Seahawks cap and leather gloves, although whenever it gets really cold, far below zero, he dons mittens crafted of polar bear fur. Sunglasses are a necessity to avoid sun blindness. Charlie is fifty-eight years old and wrinkle-free, despite long days spent in the polar sun. "The reason I look so young is that half my life I've been frozen," he says, only half joking. Every March, sometimes when temperatures sink to 60 degrees below zero, he heads out alone on the ice to hunt wolf and wolverine. Like all subsistence hunters, his life is tuned to the cycles of the animals. In summer, after the bowhead hunt, he hunts walrus and bearded seal for dried meat and boat skins. In the fall, he switches to caribou, and there's another bowhead hunt when the whales head south. He catches about two-thirds of the meat his family eats. Once a week, he buys beef at the grocery store. Every two weeks, chicken. "Most of what we eat is what we catch," he says. "This is what we need to survive." Charlie, an environmental consultant, has taken his usual spring leave from his job, a "subsistence leave," without pay, granted to hunters. The oil industry has been good to Barrow, bringing relatively high-paying jobs like his. But, Charlie says, "We're rich in culture, not money. When the (oil) money's gone, we'll still be doing this." On vacations, he visits Seattle to watch the Sonics play or sometimes vacations in balmy Hawaii, but he will never leave Barrow for longer than a week or two. "I have no business in other people's worlds. This is where I will stay. If it's too warm, I'll melt and wrinkles will come."

A duck hunter fires a shot in the distance. The radio crackles. His cell phone rings. "OK," he says, hanging up. "86 to 80. The Lakers won."

As captain, his mind is constantly working, planning, preparing, judging. As soon as one hunting season ends, he plans the next one. Recruitment "is always a struggle," he says. His crew finally arrives. Two are his nephews, only fifteen and sixteen years old, tenth graders. Every April, they leave their Gameboys at home, trading them in for ice picks to break the trail for the captain. The boys get school credit for their time hunting by keeping a log of currents and wind conditions, turning it into an experiment mixing culture and science. Hunting is considered critical to their education at Barrow High School, home of the Whalers. No child is left behind. Even those as young as eight help run errands and keep the campsite clean. Charlie and his crew head out on three snowmobiles, one dragging the skin boat, one hauling a sledge packed with tools, a stove, a shovel, an ax, a rifle, a propane tank. No tent. "We'll do without," Charlie says. The crew arrives at the edge of the ice at 2:30 P.M. They are standing on first-year ice, a few feet thick, strong enough to hold a fifty-ton bowhead.

Charlie checks its stability with a long, sharp tool that resembles a giant ice pick. Then he sets a compass down on the ice. "If it starts moving, we're in trouble," he says. Once when he was eleven years old, he got trapped on broken ice with his grandfather. They were there for two days before they were rescued. A couple years ago, he and William were onshore when they saw a chunk of ice floating by with a duck hunter sitting on it. They had watched that iceberg float by for several days and had no idea anyone was stranded on it until then.

The crew unloads the boat, rifles, and harpoon guns. This is home for a while. Charlie watches the water for the broad, blue-black back of a bowhead. "They're on the move over here," he says. The crew pushes Charlie's umiak out to the edge of the ice. The sea here is a thin sheath of slush, what the hunters call "new ice." About one hundred feet out, it is an open current called a lead, moving fast, blown by the wind. That is where the whales are and that is where they need to be, but to get there, they must maneuver through the new ice. Paddling a boat through slush is like trail-building on the frozen ocean—slow and tedious. The boat wobbles. They move only about fifty feet before they get stuck. They rock the boat like a cradle, trying to loosen the sludge. "Rock 'n' roll," Charlie says. Finally, they return to shore. The slush is too soft to walk on, too hard to row on, so they are stuck here, waiting for the current to move it offshore so they can reach the open water and the whales.

They wait. Again.

"This is a whaler's life," Charlie says. He takes out a whitefish and cuts off a slice. "Frozen sushi," he says. They bring out cinnamon rolls and bear claws (the pastry variety) and settle in for the wait. He calls his wife to tell her they have reached the ice and they are waiting for a lead to open. How many thousands of hours has he spent like this over the past fifty seasons? It seems a lot of effort just to gather food. But this is food that will provide hundreds, perhaps thousands, of meals. His job, he says, "is to make sure no one goes hungry." Last year, he caught a twenty-eight-footer, a young one, which means its meat is sweet, and he wound up having to buy five hundred paper plates for the big spring feast, then store the rest for Thanksgiving and Christmas, when he distributed three or four more tons of meat. He remembers that the bowhead came toward his boat, twenty feet from the shoreline, and swam right up to his harpoon, seemingly showing him the "kill spot" on the back of his neck. When the whale was dead, a runner took his family flag back to his wife, who hung it on a pole outside their home. It is his grandfather's flag—red, white, and blue with diamonds and a number "2" for Hopson 2. His cousin is Hopson 1.

"There's a whale out there," Charlie says. He sees its blow, a fine, V-shaped mist. They helplessly watch it. So close, yet so far.

He radios the Leavitt crew nearby. The ice is the same there. "It's lousy whaling conditions," he says. "What the hell am I doing here today? I should be at home watching TV. Half an hour more and we'll head back. Conditions won't improve with these southeast winds. It's going to bring the ice in more."

Charlie's father, Eben Hopson, wasn't a whaling captain. He was a legislator, considered one of the great historic political leaders of the north. When he was fifteen years old, Eben wrote to the commissioner of Indian Affairs in Washington, DC, to complain about the school principal's use of unpaid student labor on public works projects. When the letter was forwarded to the principal in Barrow, Eben was prevented from boarding a ship to travel to a boarding high school. Denied education, he became committed to protecting the rights of the Inupiat, and served on the Alaska state senate and as a special assistant to the governor. In 1971, he helped secure enactment of a land claims act that awarded Alaskan village corporations nearly $1 billion and title to roughly 40 million acres. Before his father's work, the world "never recognized us. We were frozen in the Arctic, forgotten," Charlie says. Eben Hopson was mayor of Barrow when the IWC imposed its bowhead ban in 1977, and his father's work to restore the hunt left an indelible impression on his oldest son, Charlie. Eben, who dreamed of the Eskimo people as one giant nation together, galvanized the people of the Arctic, forming the Inuit Circumpolar Conference, an international group that works to further the rights of indigenous peoples and maintain control over their natural resources. He died when he was about Charlie's age, fifty-seven. "My father had big dreams," Charlie says, "and they all came together."

While his father never had much time for whaling, his grandfather was a captain, and Charlie learned the skills from him. But like his father, Charlie is half poet, half politician. He ponders the Inupiat way of life while sitting on a sealskin, waiting for the young ice to move south with the current. "We're people of the ice," he says. "The bowhead and us, we live together. We were made for each other. It is us against the monster—the biggest mammal he's got. God put us up here to see if we could do it—and we can. He made a good choice, putting us here." In a poem entitled "Bowhead," in homage to the whale, he wrote that the destinies of the whale and the Inupiat are forever intertwined.

At this particular moment, his destiny is far offshore, out of reach. They set up a windbreak and his brother, William, decides to make coffee. He gets the propane out and a can of Hills Bros. Then William readies his weapons and explosives, making sure they are dry and clean. The harpoon, armed with a "bomb" and weighing about thirty pounds, is heaved at a whale by a single man, to be embedded a foot and a half into the whale's flesh and blubber. "A thousand bucks apiece," Charlie says about the exploding harpoon, which is mandated by the IWC to ensure that a whale dies quickly. Only once the harpoon hits the whale can they use a shoulder gun. The target is a square-foot patch on the whale's spine, about eighteen inches behind its eye. If the harpoon doesn't strike there, the whale disappears underwater and keeps going, dying at sea—what is known as a bad hit. Everything—weeks of trail-building, days and nights of waiting—culminates in the five seconds that a harpooner has to aim and fire. Three strikes and he's out. If the harpooner misses that many times, the boat will probably be bumped back inland by the whale.

An hour later, little has changed. "If the water doesn't come out, we're going to pack up and go home and go to sleep," Charlie says, annoyed. "I'm going to be pissed. I'm too old for this kind of work."

Three young men from a nearby crew, Charlie's cousin Harry Brower's crew, arrive to help. The eight men climb onto the boat, Charlie supervising at the stern. In unison, they paddle, sloshing through the slush an inch at a time. After half an hour, they've moved barely a few yards. They stop at 9 P.M.

"Aren't we the sorriest half-ass whalers?" one crew member asks. No one answers him.

"Let's go home," Charlie says. Sometimes life in Barrow is a bit like rowing through slush: The more you struggle, the less you gain. Charlie drives his snowmobile home. By the time he is off the ice and back onto a dirt street, it's a little past midnight. The sun blazes in the sky, low but bright, just a bit above the horizon.

The next day, warm winds blow up from the west, just as Charlie had feared. It is over 30 degrees Fahrenheit and the ice is breaking apart, too early this year, in stark contrast with the year earlier, when the open water stayed within reach for weeks and twenty whales were landed. The ice is dangerous, Charlie says. "Only a few whalers are out today. Brave ones." He stays ashore. If the warm spell continues, he will have to move his umiak. Days pass, and people in Barrow peel off their gloves and hats, wearing only shirtsleeves as temperatures soar. Slush replaces ice, mud replaces slush. Reluctantly, Charlie retrieves his umiak from the ice but some foolhardy hunters refuse to give up, misjudging the dangers of the winds. A day after Charlie ends his own season, fifty-eight of them are stranded on broken ice, drifting off to sea. They and their snowmobiles have to be airlifted out by helicopter. The ice broke close to the spot where Hopson had camped. It was one of the worst spring hunts Charlie could remember in half a century. There would be only one *nalukataq* that summer.

Before the month of May is over, there is more distressing news, reaching Barrow from Tokyo. The International Whaling Commission, meeting in Japan, voted to ban the Alaskan hunts beginning next year. Japan claimed its people had aboriginal rights and requested to start its own whale hunts yet the United States and other nations refused, prompting Japan to lead a movement to curtail other hunts. The retaliatory move was short-lived, however. Alaskans mobilized their forces, and a new vote in the fall renewed the bowhead hunts.

By the next spring, Charlie was back on the ice again, awaiting his destiny.

Bowhead
by Charlie Hopson

Silently, I move toward destiny.
Quietly, you, Inupiat, await my destiny.

I can hear you as I move under the ice.
I can see you as I surface.

Together we wait.
Both know what the other thinks.

Although we live in different worlds,
We exist for each other.

I move toward you, Inupiat,
Because it is my destiny.

You wait for me,
Because I am your destiny.

Quietly, I approach you.
Silently you move toward me.

I give you your culture.
I give you myself, Bowhead Whale.

PART II

SCIENTISTS SEEKING ORDER OUT OF CHAOS

Chapter 7

Fear Is Toxic, Too:
Communicating Risk
to Canada's Inuit

Eric Dewailly sits huddled in his office on the outskirts of Québec City, surrounded by stacks of medical journals and reports. After nearly twenty years of research, and scores of scientific articles bearing his name, he still agonizes over the Arctic dilemma, by far the biggest challenge of his medical career. It is much easier, he says with a sigh, to stop a disease outbreak than to eliminate a contaminant. This reinforces his belief, formed early in medical school, that prevention is key to protecting public health. "It is better to prevent than to cure. Logical, no?" he says in English with a heavy French accent. But Dewailly is painfully aware that it is too late to prevent contamination of the Arctic, and unlike many health threats, there is no vaccine, no cure— and there never will be. The best that he can do, as a leading environmental health expert in the province of Québec, is to give the Inuit some wise advice. But, to this day, he remains torn over what prescription to give. Sometimes he wonders if it would be best to leave the Inuit alone, telling them nothing, rather than take a chance by prescribing changes in their diet that might cause irrevocable harm.

When it comes to protecting Arctic people from contaminants, no other country has tried so hard, agonized so much and stumbled so many times as Canada. For two decades, Canadian scientists and public health researchers have sampled the breast milk of women, tested every animal species ever eaten, probed the bodies of babies, and wrestled with dietary programs—yet the nation and its aboriginal people remain mostly paralyzed over what path to take: Should the Inuit eat less of their traditional foods, or more of them? Since the mid-1980s, a wide chasm has grown between what scientists say and what the 56,000 indigenous people of the Canadian Arctic hear, fueling fears and misunderstandings. In the clash of these two worlds—science and traditional knowledge—comprehension has been a casualty. Are public health officials frightening the Inuit or educating them? Or are they just confusing them so thoroughly that nobody in the Arctic bothers to listen anymore? When it comes to toxic threats, the difference between scaring people and boring them is a fine line.

"Really what we've done is raise awareness but not raise comprehension," says Christopher Furgal, a researcher at Québec's Laval University who specializes in health risk communication for aboriginal people. "We've learned how to scare people but not how to inform them."

Canadian government leaders and public health officials have failed to find a way to refine their message that resonates with the traditional cultures of the Arctic. As a result, at least three generations of Inuit children have been exposed to the chemicals with little or no advice from experts on how to reduce their exposure. Today, many of Canada's Inuit still lack even basic understanding of the pesticides and industrial compounds in their foods and in their bodies. Most are aware of the chemicals but still don't understand what they are, where they came from, how they get in their foods, and what effects they may have. Few have taken any steps to reduce the chemicals they ingest.

Part of the problem lies in semantics. In the Arctic, where recognizing the nuances of nature is key to survival, language is driven by necessity, and Inuktitut, the language of Canadian Inuit, has a rich and detailed history of providing words that Arctic inhabitants need. There are some fifty words (actually phrases, although they resemble single words) for snow and ice. *Qanniq* is falling snow. *Maujaq* is deep, soft snow. *Kinirtaq* is wet, compact snow. *Katakartanaq* is crusty snow marked by footsteps. *Uangniut* is a snowdrift made by a northwest wind. *Munnguqtuq* is compressed snow softening in spring. Yet there are no Inuktitut words for chemical or pollution or contaminant. Over the thousands of years that their culture has existed, the Inuit have had no need for words describing toxic chemicals. They have never seen soot spew from a factory smokestack or smelled the stench of diesel truck exhaust or sprayed pesticides on crops. Out of lack of a better choice, Canadian health officials have called the chemicals in native foods *sukkunartuq*—something that destroys or brings about something bad. But use of the word has made the contaminants seem lethal and mysterious, even supernatural, leaving the Inuit confused and fearful.

Communication between health officials and the Inuit has been so poorly handled that it has caused extreme psychological distress, according to the Inuit Tapiriit Kanatami, an organization representing the Canadian Inuit. Researchers in the group launched a project in the mid-1990s to gauge the success of authorities' efforts to inform nine communities about contaminants. Fear, they concluded, is the most tangible and dangerous threat the toxic substances pose to Arctic people. "In every instance, there was a pervasive unease and anxiety about contaminants," the organization wrote in its 1995 report, "Communicating about Contaminants in Country Food." "Whether or not individuals are exposed to or actually ingesting injurious levels of contaminants, the threat alone leads to anxiety over risks to health, loss of familiar and staple food, loss of employment or activity, loss of confidence in the basic food source and the environment, and more generally a loss of control over one's destiny and well-being."

The late Ingmar Egede, the eloquent educator from Nuuk, Greenland, across the Davis Strait from Nunavut, believed that fear of toxic substances is worse than the chemicals themselves. "Contaminants do not affect our souls," he said. "Avoiding our foods from fear does."

Canada's public health quandary began with a decision to study the people of Broughton Island, a tiny hamlet in the Baffin region, in 1985.

Health officials were concerned that an Arctic air-raid warning radar system, a Cold War relic, might be leaking small amounts of chemicals such as PCBs. Led by Dr. David Kinloch of Health and Welfare Canada, medical technicians collected blood samples and breast milk of a few women on Broughton Island. The levels of PCBs were so high—much higher than what could have come from local military facilities—that Kinloch wanted to test more women, so the mayor of Broughton Island granted him permission. In the summer of 1988, the tests confirmed high concentrations of PCBs in breast milk.

At the same time, Québec's Eric Dewailly, in a separate project, was finding extraordinary levels of DDT, PCBs, and other chemicals in the women of a neighboring part of the Arctic, in Nunavik, the Inuktitut word for "great land." Their breast milk had concentrations up to ten times higher than milk from women in southern Québec, around Montreal. Before the data could be analyzed, and before people in the villages were notified, the discovery leaked to the press. On December 15, 1988, Toronto's *Globe and Mail* published a front-page story, quoting an Environment Canada official saying that the Inuit were so contaminated that they may have to eat beef and chicken and give up whale, seal, and walrus. The Inuit were terrified and some stopped eating their native foods, seeking the government's assistance. Virtually overnight, Arctic contaminants became a crisis for the Canadian government.

In early 1989, the department of Indian and Northern Affairs convened a meeting to discuss the findings and figure out a strategy. Scientists and health leaders met at the Chateau Laurier, a grand old hotel in the Canadian capital of Ottawa. With its marble floors, high ceilings, and spires, the opulent hotel is reminiscent of a fifteenth-century castle—far in distance as well as spirit from the icy hunting grounds of the Arctic. Aboriginal leaders came to the hotel and begged to be included in the meetings, but government officials refused. They met behind closed, locked doors at the hotel, right across the street from Canada's Parliament buildings. Kinloch traveled to Broughton

Island soon afterward to inform people there of his findings, but the presentation was too complicated for the Inuit to understand. Even a scientist who heard one of the initial presentations remembers that the information was virtually incomprehensible. The poor communication effort, combined with the secrecy at the Chateau Laurier, were a slap in the face that Canada's indigenous people, to this day, have not forgotten. People in the Baffin region were furious. For more than a century, they had been neglected or mistreated, similar to the early plight of American Indians, and the chemical crisis added fuel to their already growing movement to win independence from the Canadian government. David Stone, now director of northern contaminants research at Canada's Department of Indian and Northern Affairs, attended the Chateau Laurier meeting and remembers the furor well. "It was a disaster," Stone says. "The good news is that we learned our lesson. We learned it painfully but we learned it very well."

Suzanne Bruneau had lived for several years in the Arctic village of Kuujjuaq, studying the safety of its potable water, when the discovery of chemicals in the breast milk of Nunavik women was revealed in the late 1980s. Eric Dewailly came to the village looking for her. He asked if she would stay in the north to help with his study of the contaminants. She accepted, and soon, mothers began turning to her for advice, some so worried that they had stopped nursing their babies and eating their native foods. Bruneau didn't speak their language so, through an interpreter, she encouraged them to follow their traditional ways. She didn't know much about contaminants, but she did know that babies would die if their mothers stopped breast-feeding. The Inuit could not afford to buy milk and had no access to infant formula. She recalled one horrifying story, recounted by a doctor, of a woman who had fed Coffee-mate to her adopted baby, not because breast milk was contaminated but because she didn't have any and couldn't afford infant formula. Bruneau didn't want to scare people and risk deaths from malnutrition. But she felt uncomfortable sending such an ambivalent, confusing message about the chemicals. She needed help.

In Ottawa, Canada's leading medical experts were no more prepared for the onslaught of panic sweeping through the Arctic. Health Canada, the nation's public health agency, was paralyzed with indecision. The Nunavik and Baffin data clearly showed that most Inuit were exceeding the agency's "tolerable daily intake levels" for toxic contaminants. If the agency were to adhere to its own national policies, it would have to issue warnings to the Inuit to stop eating their traditional foods. But public health officials had never encountered a problem like this before, where the contaminated foods were so vital to a society's health, culture, and economy. On the one hand, it seemed dangerous to advise people not to nurse their babies and eat their foods when alternatives were unavailable. On the other hand, if they ignored their own toxic guidelines when it came to the Inuit, wouldn't that be discriminatory? Could they follow one health policy for white people and another for the Inuit? And if they were going to ignore the guidelines whenever they were too difficult to implement, why have them at all? Scientists could not reach a consensus on whether Inuit foods were safe to eat. Some wanted to advise the Inuit not to eat them, but Dewailly thought it was dangerous to give such sweeping advice about food that was so important nutritionally and culturally. Canada already was immersed in divisive issues about the treatment of its native people, and this one was particularly sensitive. Health Canada officials decided they needed more information on the health effects before they could determine what advice to give. The Québec Ministry of Health turned to Dewailly for help.

Dewailly was mostly worried about the babies of Nunavik. He knew the chemicals could pass right through a mother's uterus and enter the body of a fetus. He had wondered for years why Nunavik children had an extremely high incidence of infectious diseases—in particular, bronchial infections, meningitis, and ear infections called otitis. About one-quarter of Nunavik children had such frequent and severe ear infections that they suffered hearing loss. Reading up on PCBs and organochlorine pesticides, he learned that they were known to harm the immune systems of animals. Could they be causing the Nunavik epidemics? He mounted an investigation to determine whether babies exposed to high concentrations of contaminants suffered more illnesses. It began with a Nunavik baby born in July 1989, and involved 170 more born over the next year. Dewailly's team followed them through their first year of life, documenting their illnesses and probing their immune cells. They concluded that ear infections increased with prenatal exposure to the pesticides DDT, hexachlorobenzene, and dieldrin. It was the first evidence that the Arctic's legacy of contaminants could be harming native people.

Their next challenge was to explain the findings to the people of Nunavik. Public health officials realized, after their mistake on Broughton Island, that in-person communication had to be their top priority. They didn't want people there to learn about it from the media again. Dewailly and other scientists prepared pie charts and bar graphs, flew to Nunavik villages, and organized community meetings, where they talked in detail about organochlorines and immune suppression and disease rates. But after a series of presentations using translators, the experts realized that they kept getting the same questions over and over, proof that even their most basic points were failing to come across. What are these chemicals? people asked. How did they get here? Can we see them or taste them? The cultural barriers seemed insurmountable. Language was one obstacle—the Inuit spoke no French or English and the scientists spoke no Inuktitut. Scientific illiteracy was another. Medical and scientific terms simply did not translate into the aboriginal languages. Most Inuit over the age of forty have had no schooling, so they cannot read and they do not understand percentages or charts. Instead, they believe in the power of traditional knowledge—what they see for themselves or learn from their elders' lore. They can't see the contamination, can't touch it, can't smell it, and there aren't even words to describe it, so many refuse to believe it exists.

Although they are aware of diseases like trichinosis that pose an invisible threat in their food, they learned of them not through outside health experts, but through their elders, who witnessed people falling ill and dying. With contaminants, there is no traditional knowledge to pass from generation to generation. Outsiders are often distrusted, particularly government authorities, since many remember Canada's century-long efforts to erase their heritage and assimilate them into modern society.

When a planeload of scientists flies to their remote villages, the Inuit become alarmed, even if the message is not intended to be alarming. They notice that the doctors and scientists voice many doubts when they describe the dangers of the chemicals. They can hear the uncertainty in their voices and see it on their faces even if they don't understand the exact words. "People want a simple answer to a complex question but to give them a simple answer is, in some ways, unethical," Furgal says. Scientists, by their very nature, often dwell on gaps in their knowledge, and when it comes to toxic substances, they send mixed messages, essentially telling the Inuit, "We know your foods are healthy, but they might be dangerous, too."

Because of what Furgal calls the "culture of contaminants" that began so unfortunately on Broughton Island, the Inuit are skeptical of the government's advice. The furor over contaminants prompted Canada in 1991 to create the Northern Contaminants Program to investigate toxic substances, with the mandate to empower native communities from the onset. Soon afterward, the Baffin region won its independence. Canada's parliament created a new territorial government. Nunavut—"our land" in Inuktitut—was born, home to 21,000 indigenous people, and the Inuit vowed to never again allow anyone to control their destiny.

The national government has now stepped aside when it comes to issuing food advisories in the north. Instead, the advisories are created locally, tailor-made for each region. In Nunavik, a committee of public health officials and Inuit leaders agree on dietary advisories. "There's no big brother now," says Eric Loring of the Inuit Tapiriit Kanatami.

A lone hunter walks along a frozen fjord in the Thule region of Greenland, the northernmost civilization on Earth. Inuit there are the closest to the archetype of traditional polar life, hunting narwhal, seals, seabirds, beluga, and walrus.

As spring narwhal hunting season begins in early June, the Kristiansen brothers set up camp on the ice about thirty-five miles from their village of Qaanaaq, in northwestern Greenland.

Gedion Kristiansen of Qaanaaq drives his sledge toward ancestral hunting grounds about eight hundred miles from the North Pole.

Mamarut Kristiansen untangles his team of sledge dogs while on a hunting trip in northern Greenland.

The village of Qaanaaq, Greenland, is home to about six hundred Inuit, who live in Scandinavian-style prefabricated houses provided by the Danish government.

Mamarut Kristiansen says patience is a critical virtue for Inuit who hunt narwhal, a reclusive whale with a spiraling tusk, from a sealskin kayak.

Qaanaaq hunter Gedion Kristiansen wears a fur-lined parka as temperatures dip on a June hunting trip.

Her four-month-old cubs nestled beside her, a polar bear mother sleeps on the ice in Svalbard, Norway, after being tranquilized by scientists who are measuring contaminants.

Scientists Magnus Andersen (left) and Andrew Derocher (right) take fat and blood samples from a bear and two cubs on Svalbard, where bears are highly contaminated.

A lone male bear ambles across the ice on a spring day in Spitsbergen, part of Norway's Svalbard Archipelago, about six hundred miles from the North Pole.

Scientist Andrew Derocher extracts a female polar bear's tooth, a useless premolar, which will be used to estimate her age.

Derocher carries a four-month-old cub, which weighs about fifty pounds.

Charlie Hopson's whaling crew tries to row through thick slush to reach bowhead whales off Barrow, Alaska.

Young hunters gather on the ice near Barrow Point, Alaska, to watch for herds of giant bowhead whales, which can reach fifty feet long.

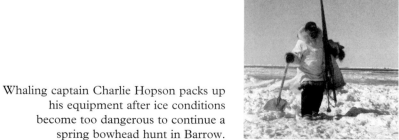

Whaling captain Charlie Hopson packs up his equipment after ice conditions become too dangerous to continue a spring bowhead hunt in Barrow.

Inupiat in Barrow gather for a *nalukataq*, or spring festival, to celebrate the bowhead hunt. Whaling crew members jump on a blanket toss, a trampoline created from a sealskin removed from a captain's boat.

Crew members slice *maktak*—skin and blubber—
off a bowhead whale's flipper to serve to the community.

Every family gathered at the spring festival in Barrow brings home
enough chunks of bowhead meat and blubber to fill an ice chest.

Inuit in Greenland hunt narwhal the traditional way—one man in a sealskin kayak, armed with a harpoon, pitted against a one-ton whale. Gedion Kristiansen is one of the best hunters in his village of Qaanaaq.

Their goal is to keep the message simple. Instead of holding community meetings, which were always poorly attended, the scientists are appearing on local radio stations, answering call-in questions, and displaying brief, colorful posters in markets and doctors' offices. Still, efforts to simplify sometimes have gone awry in unpredictable ways. Bruneau tried using drawings of human beings to display the results of each individual's test for contaminants. She knew she couldn't use pie charts or percentages. So she drew a line across the picture of one body to show one person's contamination level and another line across another picture to show the village's average. Bruneau thought it would be a simple way to help people compare how their own contamination compared with others. But she soon discovered that some people looked at a line drawn through their knees and thought that meant they would have problems walking.

By 1997, the public health officials and Inuit leaders developed an official message: The benefits of the native diet outweigh the risks. Don't worry, they said. Just keep doing what you are doing. Eat your foods. But there was a visible rift in the scientific community. Some thought they were going overboard with the "everything's OK" message, that the risks were being downplayed and the science watered down to push the status quo. Others, including Dewailly, thought it was the only rational message to avoid overreaction. If he told people to reduce their consumption of beluga, the main source of the chemicals in Nunavik residents, their rates of heart disease, cancer or malnutrition could increase because beluga is full of healthy fatty acids. He decided against any advice that would be construed as a recommendation against native foods or breast milk.

It turned out that the fears that people would shun their foods were largely unfounded. "Heightened awareness of potential contamination does not on its own cause the majority of people to cease harvesting activities or reduce consumption, at least for any length of time," the Inuit Tapiriit Kanatami says in its report. In the 1990s, 42 percent of women questioned in Nunavik increased consumption of traditional foods while pregnant. The main reason they cited was that it was good

119

for their babies. Of the 12 percent who ate less native foods while pregnant, only one of 121 said the reason was to avoid contaminants. At a 2002 focus group where residents expressed their worries about health issues, Bruneau says contaminants never came up. There is no evidence that people have changed their diet because of the chemicals. They ask a lot of questions, are upset about it, but then go home and keep living the way they have been for generations.

Had the "Don't Worry, Be Happy" message gone too far? By 2003, new evidence of health effects began to erode the scientists' confidence in their optimistic message.

Dewailly wondered if the food contamination was harming not just the immune systems of Nunavik babies but their mental abilities. He knew, from a landmark 1990 study by Michigan scientists, that children whose mothers ate a lot of PCB-tainted fish from Lake Michigan had reduced memories and cognitive skills, and he wondered if the same thing was happening in Nunavik, where contaminant levels were far higher. In 1993, he contacted Gina Muckle, a developmental psychologist at Laval University. She knew nothing about Inuit culture, but she was experienced with social determinants in child development, such as family violence and alcoholism. She recognized from the onset that if contaminants had any effects, they would be subtle. In 1995, Muckle began to assemble Nunavik babies in a project that was supposed to take four years but lasted seven years because of difficulties conducting tests in tiny, remote villages. The babies were given tests designed by neuropsychologists to gauge their intelligence. They looked at pictures to see how quickly they could recognize faces. They tried to find a hidden toy. Muckle was seeking answers to questions that would help decide whether Inuit foods were safe to eat: Were the benefits of fatty acids in breast milk stronger than the effects of the PCBs? In 2002 she had the results: Despite the benefits of breast milk, the PCBs seemed to be affecting the skills of the babies. In studies testing their ability to recognize faces, babies with low PCB exposure succeeded in recognizing them 55 percent of the time compared with 50 percent for babies with high exposure. She also found that the PCBs lowered their birth weight. Babies with high exposure weighed six ounces less, on average, than those with low exposure. The differences were subtle, Muckle says, but similar in scope to the effects of a mother who drank moderate amounts of alcohol while pregnant. Scientists are now conducting other neurological tests on Inuit babies, from reaction time in tapping keys to vocabulary tests.

When it comes to effects on childhood diseases, the picture isn't any clearer than it was a decade ago. Dewailly's team has examined the children's infectious diseases in more detail, yet their latest findings aren't any more conclusive than the earlier ones. They found a "slight tendency" of children with higher exposure to the toxic substances to suffer more diseases in their first six months of life, Dewailly says. The findings, however, remain questionable, as they involved only 199 children. What is clear is that the babies of Nunavik are still suffering an inordinate number of infections. "I continue to feel that all the issues of infection in the first year of life are more important than a slight neurological delay," Dewailly says. "I don't know what it means to have a slight delay in a finger-tapping test but I know what it means to have pneumonia in the first year of life."

After months of debate about how to react to the new findings, the Nunavik health committee, consisting of Inuit leaders as well as Québec medical experts, decided that it must shift its message. The goal was to reduce the exposure of women of childbearing age to contaminants in their food without reducing its nutritional benefits. In 2003, the health committee decided to focus on telling people what they *should* eat rather than telling them what they *should not* eat. They concluded that Arctic hunters may have to forsake certain prey and learn to rely on less-contaminated ones to preserve their health as well as their culture. Women were advised to eat Arctic char, a tasty, popular fish that has low contaminants but high amounts of beneficial fatty acids, and, in a pilot program, free char was distributed in three communities. The notion is that if the Inuit eat more char they will eat less beluga, the source of two-thirds of the PCBs in Nunavik residents. Telling people to add char to their diet is similar to telling people in lower latitudes to eat more fruits and vegetables. It is a positive message—and it avoids all confusing and divisive talk of contaminants—giving people workable options for changing their diet. "Eat more fish and less mammals. The advice is that simple," Dewailly says. Still, he wavers. He remains confident that the benefits of the food and breast milk outweigh the risks, although he now acknowledges the need to reduce exposure and supports the program to encourage consumption of char. Nevertheless, he has lingering doubts because char isn't as nutritious as beluga. "I'm not absolutely convinced we should restrict beluga fat," he says.

Balancing risks and benefits of food is not an easy task, which explains all this wavering and hand-wringing by Canadian health officials. Pierre Ayotte, a Canadian expert in human toxicology, says it is more of a judgment call than a science. "People talk a lot about risks and benefits. Easier said than done. It's a little bit of an illusion to say we'll have models to govern risks and benefits." Much more is known about the benefits of Inuit foods than the risks. Ounce for ounce, few diets are as healthy when it comes to nutrients but as risky when it comes to contaminants. While oily fishes such as salmon and char contain lower doses of chemicals and relatively high doses of omega-3, even they cannot compare to the nutrients in whale and seal.

Arctic people have few healthy alternatives even if they do want to reduce their exposure to contaminants. Jose Kusugak, president of Inuit Tapiriit Kanatami, the organization representing Canadian Inuit, says he can buy "lame lettuce" and "really old oranges" and "dried-up apples" in his Nunavut village or he can eat fresh and nutritious beluga, walrus, fish, and caribou. "There is really no alternative," he says. Dr. Harriet Kuhnlein, founding director of McGill University's CINE (Centre for Indigenous Peoples' Nutrition and Environment) in Montreal, reports that 30 percent of the dietary energy of Inuit living in villages comes from traditional foods, an amount that would be hard to replace with expensive imported foods. Fish and sea mammals are a daily part of their diet. Nevertheless, food is often scarce for the Inuit. A 2004 study by the Nunavik Regional Board of Health and Social Services in the village of Kuujjuaq showed that more than half the people there live in poverty and 68 percent were at risk of not getting enough food. For children, the figure rose to 74 percent. Low- and middle-income households experienced at least one episode a year when they didn't have enough to eat.

"Unless nutrient supplements are routinely used, a decrease in use of sea mammal foods will place some individuals at health risk from nutrient deficiency," says a CINE study, published in 2000, called "Assessment of Dietary Benefit/Risk in Inuit Communities." In the survey of indigenous people, "respondents ranked traditional food as significantly more important than market food as healthy for children, healthy for pregnant and breastfeeding women, tasty and important to community life."

As some Inuit move away from their traditional foods and eat a more American or European diet of processed items loaded with simple carbohydrates and sugar, rates of obesity, heart disease, and diabetes are climbing in some Arctic populations. On the days that the Inuit eat no traditional foods, they take in considerably less protein, iron, vitamin A, and omega-3 and more carbohydrates, saturated fats, and sugar. Not only are young people in some Arctic towns eating less nutritious foods, but they are also hunting less—so they are more sedentary. "We know people are eating less and less traditional food," says Peter Bjerregaard, head of Greenland research at Denmark's National Institute of Public Health. "Old people eat a lot of traditional food and young people very little."

Reduced consumption not only harms their physical health, but apparently their mental health, too. Psychologists and neurobiologists at the Institute of Arctic Biology at the University of Alaska–Fairbanks reported in 2003 that "the mental health of circumpolar peoples has also declined substantially during the same time period" as the increased health problems. "The decline in mental health is characterized by increased rates of depression, seasonal affective disorder, anxiety, and suicide, that now often occur at higher rates than in lower-latitude populations," the scientists wrote in the *International Journal of Circumpolar Health*. "Studies in non-circumpolar peoples have shown that diet can have profound effects on neuronal and brain development, function, and health. Therefore, we hypothesize that diet is an important risk factor for mental health in circumpolar peoples." The suicide rate for adolescent Inuit is at least ten times higher than in other cultures and its homicide rate compares to those in major U.S. cities, Bjerregaard says.

In the cultures of the Arctic, as with many aboriginal peoples, the lessons and traditions of the past are far more important and enduring than the realities of the present. Bucknell University sociocultural anthropologist Edmund Searles, who researched the bond between indigenous identity and food in Nunavut, wrote that sharing native foods has framed the historic as well as the modern identities and social experiences of the Inuit, and that the contaminants are already altering their traditions and attitudes toward food. "Although Inuit continue to view the sharing of seal meat as a vital source of family and community wellness and . . . individual health, these practices have taken on new meanings as hunters confront the challenges of modernity, such as contaminants in the food chain, more stringent regulations on hunting, and increasingly expensive technologies. As a result, Inuit have been forced to reconsider the place of hunting, the value of seal meat, and the centrality of sharing in Inuit society," Searles wrote.

If hunting is discouraged, people quickly lose their traditional knowledge about the environment and their hunting skills, as well as material items such as tools and clothing, says Robert Wheelersburg, an anthropologist at Elizabethtown College who specializes in Arctic cultures. Even their language would suffer. Inuit dialects are steeped in the subtleties of nature that their national languages—English, Danish, and French—ignore. Wheelersburg says the most important damage would be to Inuit "values and attitudes." In the subsistence economy, people share prey among neighbors and relatives, even strangers. The best hunters are also leaders in the village, and they share their wealth. If the Inuit switch to a cash society, that communal generosity would disappear. "It's more than the food you are changing," Wheelersburg says. "It's the actual catching and hunting of it that really generates the cultural characteristics." Even skipping one generation would impair hunting skills, he says, and "once they are lost, I don't see how you can regenerate them."

Communication efforts have improved greatly in recent years but Furgal says he is still frustrated by the inability to provide the Inuit the information they need to make their own choices. The science of risk related to contaminants has emerged from around the world, and scientists are now skilled at weighing the dangers of chemicals. But little research has focused on learning how to tailor a message to specific cultures or understand its outcome, and without such efforts, Furgal says, how can they ever really know whether they are frightening the Inuit, boring them, or something else entirely?

Sixteen years after publishing his first results, Dewailly still oversees epidemiological studies of the children of Nunavik, and he still struggles over what dietary advice to give people. Certainty evades his prescriptions for the Arctic—and always will. As a doctor, Dewailly is accustomed to explaining well-known risks and well-known benefits. With botulism, for example, doctors know the agent, they know why it kills, know how to advise people to avoid it. With a prescription like aspirin for headaches, they know it works without serious side effects in adults. Other public health dilemmas are a bit trickier. Increasing water temperature in public plumbing systems, for example, kills bacteria that causes Legionnaires' disease. But it also causes burns. At least in cases like that, Dewailly can make a reasoned decision by comparing disease rates to burn rates and deciding which is more dangerous. With contaminants, such clear data to compare risk and benefits remain virtually nonexistent.

Making matters worse, the public's view of many environmental risks is distorted. People often worry about things they shouldn't and dismiss things they should worry about. Dewailly says it comes down to an imbalance between control and knowledge. When people have control and knowledge about a risk they face, people and experts agree. An example is car accidents—people will buckle up because it saves lives. But when people have no control over a problem and no knowledge—and this is where contaminants fall—there is a huge discrepancy between what people and experts believe. In the face of such uncertainty and distortions, some experts won't issue any advice at all. Canadians learned the hard way that information comes out anyway, so people should learn from the experts. Nevertheless, Dewailly believes that 90 percent of the data shouldn't be communicated. "It's too much information," he says. He worries about how his own data are used, and he strongly believes that people must be empowered to make their own decisions, without undue influence from experts like him. "I have learned to be very humble," he says.

One day in 2002, on the bulletin board outside his office, Dewailly posted an article written by an Ontario scientist calling preventive medicine "arrogant," "presumptuous," and "overbearing." The author reminded public health officials to make sure their prescriptions never violate medicine's most basic of creeds: First, do no harm. Wherever Dewailly goes, his conscience is haunted by the words of his mentor in medical school: "If you can't help, at least don't harm." It pains him that he cannot say, with certainty, that he has lived up to that pledge in the Arctic.

Scientists throughout the circumpolar north know full well that there will never really be answers, just more questions. But they keep searching for the truth, mounting extraordinary efforts to search for effects of contaminants. Some disturbing clues have been found in the most unlikely of places—inside the brains of children in the Faroe Islands, where first graders are the subjects of one of the longest and most intensive environmental experiments ever conducted on humans.

Chapter 8

Into the Brains of Babes: Searching for Clues in Faroese Children

Bjørg knits her brow and chews her lip as she tries to navigate a mind-frazzling maze. She pushes her blond bangs from her forehead, rubbing with such vigor that she seems determined to reach deep into her brain and yank the answer out. When she succeeds, her face lights up with a wide, gap-toothed grin. When she fails, she frowns, turning to the next page of the test.

At the age of seven, Bjørg is a typical first grader in all ways but one: Her brain has been probed by scientists since she was two weeks old. Once a year, Bjørg has electrodes fastened to her head, measuring in milliseconds how rapidly her brain receives signals. She has been asked, countless times over her lifetime, to recall lists of numbers and shopping items. She has searched the crevices of her mind for words in vocabulary tests, tapped computer keys to check her reaction time, arranged and rearranged blocks in patterns, and copied drawings more times than her parents can remember. She has surrendered samples of her hair and blood since the day she was born. And she will likely continue to be examined by neuropsychologists until she reaches adolescence.

After two decades of intense scientific scrutiny, Bjørg and other school children of the Faroe Islands have become sentinels for the rest of the world, particularly inhabitants of the Arctic, warning of the dangers of mercury in seafood. The years of rigorous testing on Faroese children suggest that a mother's consumption of mercury impairs her child's intelligence in subtle ways. These findings, while controversial, show that children are the most fragile victims of the chemicals invading the far North, yet the signs of damage are so subtle that even their parents are unaware of them. Before babies are even born, the damage has already been done. The mercury passes from mother to fetus in the womb, disrupting the brain as it grows and irreversibly reducing the child's intelligence and mental skills. The most highly exposed seven-year-olds inhabiting these islands lag behind their schoolmates in some skills—particularly short-term memory, vocabulary, and attention span—by as much as a school year, losing, on average, five to six IQ points. Some of the neurological impacts appear to be long-lasting, perhaps permanent, based on tests of the children when they reached fourteen. The results, while controversial among scientists, have had worldwide repercussions, prompting the U.S. government to advise all pregnant and nursing women to limit the amounts and types of fish they eat.

Descendants of ninth-century Vikings, the Faroese are arguably the world's best mariners. More than a thousand years ago the Vikings sailed west from Norway, looking for new land, discovering a remote cluster of islands in the North Atlantic halfway between Norway and Iceland. Pirates found them, too, plundering the islands in the 1700s. Today, its fishermen catch some of the finest fish in the world. Yet the Faroese live in obscurity. Although the islands are part of the kingdom of Denmark, they are closer to Scotland than to Denmark, and most of their fellow Danes have no idea how the 47,000 Faroese people live, vaguely envisioning people in caves or on farms with no televisions or roads or electricity. Even the British, their closest neighbors, don't know where on Earth the Faroes are.

The islands rise from the Atlantic in cliffs so sheer they could have been sliced by a razor. Because they are planted squarely in the path of the Gulf Stream, their weather is moderate, with an extraordinarily narrow range of temperatures—from 37 to 51 degrees Fahrenheit, all year-round. But in winter, winds blow with hurricane force, their 100-mph gusts strong enough to knock a grown man off his feet and down the vertical face of the cliffs to be pounded by huge, rogue waves. Traditional Faroese homes are built on cobbled streets with a stone wall facing the prevailing winds and a wood wall, covered with tar, on the other side. Unlike Greenland, this land is green, its rolling hills the vivid hue of a shamrock. Life here, in the sub-Arctic, is not as austere as it is farther north. It is easier to survive here, even prosper. The ocean never freezes, and it is more generous with its bounty. The sea is everywhere, around every bend, every crook of the road, every mountain peak, lapping against the cliffs. At no time are any of the inhabitants of the eighteen islands farther than three miles from it. Even when you can't see it, when the fog is drawn over the islands like a thick curtain, the ocean leaves an indelible mark on the people and the land.

Most of what people in the Faroe Islands eat lives in the ocean, so like all the top predators of the sea, their bodies are filled with contaminants. The North Atlantic waters here are fairly clean compared with the Great Lakes or the Baltic Sea or other ocean waters that are fished. But people in the Faroes rely more on the sea for their diet and they eat high on the food web. The people here are Danes, but there is one thing that bonds them with Arctic people: They eat whales. Like narwhal in Greenland and beluga in Canada, the pilot whales they consume are predators, feasting on squid and fish, so their meat and blubber are highly contaminated with mercury. Mercury is a natural element, abundant in the Earth's crust, but levels in the environment have increased threefold since preindustrial times. The largest source of global emissions is coal-burning power plants, largely in Asia. No one knows exactly where pilot whales pick up mercury. Even the path of their migration remains largely a mystery, although they apparently travel throughout Europe's North Atlantic, swimming up to the Faroes from Spain.

Mercury moves about differently than PCBs and other persistent compounds. Mercury, says Trent University's Mackay, has "a Jekyll and Hyde act." It is transported in the air in the form of mercury vapor, a metal that is fairly benign because it does not build up in animals or people. But once in the oceans, it undergoes a complete chemical transformation. It converts into its ionic form and turns into methylmercury, which is highly toxic and accumulates easily in tissues. Animals like the pilot whale that are high on the food web are especially vulnerable to accumulating it. A piece of pilot whale contains up to one hundred times more than a piece of cod.

Other than fish and wild sheep, pilot whale is one of the few foods that does not have to be imported to the Faroe Islands, where the terrain is too rugged and the weather too harsh for crops or livestock. In a whaling tradition called *grindabod*, unique to the Faroe Islands and dating back four hundred years, pilot whales are herded toward beaches and their throats slit with long, sharp knives. Although the technique is designed to kill the whales quickly, it is an extremely bloody event. The waters offshore turn red with the blood of a hundred or more whales slaughtered simultaneously. Between 600 and 1,000 pilot whales are killed a year, and the tons of free meat and blubber are shared among thousands of people. One school of 350 whales can feed 12,000. The meat is divided in an elaborate procedure codified in law and enforced by the local sheriff. From the baby in a crib to the old person in a wheelchair, everyone gets his or her share of pilot whale.

Aware that pilot whales build up mercury in their tissues, Pál Weihe and Philippe Grandjean teamed together in the mid-1980s to figure out whether it was having any effect on the children of the islands. Passing through the placenta, mercury in high levels can attack the cells of a brain as it grows, dispatching them to the wrong places and disrupting the architecture of the brain—how it is constructed and organized. It seems to target the brainstem as well as other areas of the growing brain.

Upon examining hundreds of children, the team found "widespread effects on cerebral function" in the Faroese children at what had previously been considered a low and safe level of mercury exposure—from 1 to 10 parts per million in mothers' hair. That amount is prevalent worldwide, including the United States, particularly in coastal areas where people eat a lot of fish. The scientists looked at the Faroese mothers' hair, milk, and blood to see how much mercury was in their bodies at the time of pregnancy, and then looked for a connection to changes in the cognitive abilities of their children. What they found were subtle but real differences. While the effects are mild, the pattern is clear: When the dose of mercury a child is exposed to in the womb increases, his or her performance on intelligence tests declines.

"The bottom line is that IQ is affected," says Grandjean, an environmental science professor at both Odense University in Denmark and Harvard Medical School. "Each doubling of mercury takes you back (in intelligence) one or two months at the age of seven years, when development is particularly rapid. There's a definite risk that these kids are not capable of catching up."

Other research has shown that adults who eat a lot of seafood can experience memory lapses and other neurological effects. But it's the sons and daughters of mothers who eat a lot of seafood while pregnant who are most at risk. The central nervous system of a fetus is mercury's most vulnerable target.

Their findings are so worrisome that many colleagues have subjected Weihe and Grandjean to endless questions and speculation. Could it be lead, not mercury, that caused the damage? Could mercury just be masking socioeconomic factors, such as poverty or lack of nutrition, that were the real reason for the changes in the children? Reviewers knew the findings would cause shock waves globally, so they wanted them clear and convincing. Grandjean and Weihe spent years backing up their findings, eliminating as many confounding factors as they could before the first results were finally published in a scientific journal in 1997. Grandjean warns that scientists cannot control children as they do lab rats. A human study like theirs, particularly on children, generates scientific controversy, and always will. But after expanding their research to three groups totaling more than

1,700 children, the two scientists are convinced that mercury is harming the children.

Weihe, a native of the Faroe Islands and chief physician for the Faroese Hospital System, had hoped that the damage would disappear as the children aged. After all, his ultimate motive has been to help his people. But he was dismayed to discover that the effects persist throughout the school years, and most likely even longer, into adulthood. More than 850 of the children in the original 1986 group were tested until they reached age fourteen, and initial results from the teens, published in 2004 in the *Journal of Pediatrics,* show that at least some of the damage is irreversible. As was the case when they were seven, their brains at the age of fourteen responded less quickly to signals from their ears if they had been exposed to elevated levels of mercury in the womb. Also, the seven-year-olds and teens with high exposure had neurological changes in their cardiovascular systems that made them less able to maintain a normal heart rate, a condition that increases the risk of heart attacks later in life.

"Here we have actual effects on humans," Weihe says, "not based on factory accidents but on humans who are living their normal way, as they always have."

To the blind eye of science, Bjørg is subject #24 in cohort 2. Allan, born thirteen days later, is #2. Like most of his classmates, Allan's hair is the color of straw and his eyes a pale crystalline blue. He lives in a village of three hundred, where his mother is an office clerk and his father is a fish farmer, a lucrative profession on the islands. Bjørg is tall and long-legged, with a broad smile that comes easily and shoulder-length blond hair and bangs brushing her eyes. She wears expensive embroidered jeans and a striped pink-and-purple sweater. Bjørg and Allan obviously live comfortable lives. Sociologists say the Faroes' homogeneous genetics, its affluence, and low rates of family instability and alcoholism make it ideal for studying neurotoxins like mercury because such social factors can influence a child's intelligence and confound the results. Most Faroese children are from stable homes with attentive parents, first-rate medical care, and fine schooling. They live modern lives in modern houses with modern cars, televisions, computers, and video games.

With his mother by his side at Weihe's mercury clinic in Tórshavn, the capital of the Faroes, Allan listens intently to the instructions of neuropsychologist Frodi Debes. Allan's tennis shoes swing in the chair, only the tips of his toes reaching the floor—a reminder that subject #2 is a mere seven-year-old. Since infancy, the 180 children of cohort 2 have undergone tests designed by child psychologists. Each test, conducted annually on each child, measures a particular brain function. Combined, the tests last more than five hours. As excruciating as this testing is for such young subjects, it offers a detailed snapshot of how a child's neurological system functions. Comparing the results of these intelligence and neurological tests on hundreds of children gives Weihe and Grandjean a clear image of whether the consumption of pilot whale is altering their mental abilities.

Debes flashes cards of black-and-white images and asks Allan to quickly name them in his native Faroese. He sails through the first dozen or so—tree, house, scissors, flower, saw, helicopter, octopus, mushroom. He pauses at hanger—the word is a difficult one in Faroese. Even fourteen-year-olds don't know it. Allan goes on, naming wheelchair and camel, before pausing again at pretzel, an item rarely seen in the Faroes. He names more—bench, tennis racket, snail, volcano. He stops at seahorse and smiles at his mom, then shyly guesses. He continues with boot and globe, before missing others, some unknown to the Faroes—beaver, igloo, stilts, unicorn, dominoes. As the words get harder—stethoscope, scroll, asparagus—he misses many and Debes stops the test, following written guidelines to ensure uniformity. Allan scored 30 correct, a score considerably better than average among the tested children. Called the Boston Naming Test, this is a critical exam, one in which mercury exposure from pilot whales seems to have the most noticeable impact on the children. For every doubling of prenatal exposure, the tests found a two-month delay in their vocabulary skills, comparable to a 1.5-point drop in IQ. Those exposed to ten times more mercury had a seven- to eight-month delay—an entire school year.

Allan is calm, serious, and receptive, always silent until asked a question. He and most Faroese children have experienced such testing many times before. He looks forward to some of the tests, dreads others as boring. Debes recites a shopping list of fifteen food, toy, and clothing items. On the first try, Allan recalls five, slightly above average. Debes does this four more times, and each time, Allan can name more, ending with twelve out of the fifteen on the fifth try. Then Debes moves to List B—fifteen more words used as a distraction. Allan sighs. He cannot remember a single one. The examiner then asks him about the original list of words. Allan names eight, then squirms in his chair and rubs his head, trying to remember more. This test is another important one, where the studies have found a link between prenatal exposure and performance, in this case memory.

In a test of his fine motor skills, Allan is asked to stack small green blocks, copying Debes' designs, which become harder as the session progresses. He completes them all, quickly. Allan then moves on to copying drawings of shapes using a pencil. He draws them easily, meticulously, carefully, until he is asked to draw a cube. Few seven-year-olds can draw one. He is asked to recall the shapes he drew; he names seven out of sixteen. Then Debes asks him to again name the items on the shopping list he heard a half hour earlier. Allan moves on to a computer, where he is asked to tap a red key when a certain image—the silhouette of a cat—appears. Each second an animal—goat, pig, rabbit, or cat—appears. Allan never misses the cat, but he does tap the key a few times when he shouldn't have. The computer measures the time it takes for him to react, but it also counts errors, measuring attention span and reaction time, another set of skills that are impaired by mercury, according to the tests.

By the time Allan completes the first session, more than an hour has passed—and he still has four sessions to go. A technician strings a red wire across the right side of his face, a green one behind his right ear, a black one behind his left ear, a yellow one on top of his head, and a red one on the back of his head, all the while asking him about school and football to put him at ease. She secures the electrodes with a white headband. Then he lies down on the bed. The lights are turned off. He stares, silently, at a seventeen-inch screen flashing checkerboard patterns. The only sound is the humming of a computer. The examiner checks Allan's neurological signals as they appear on her screen, measuring the time it takes for an electric signal to move from his retina to the visual cortex of his brain. Next he dons headphones, listening to 65-decibel clicks, no louder than a normal conversation would be. A computer records how rapidly his brain receives messages from his ears. In these studies, for every doubling of mercury levels, electrical signals sent from the ears to the brainstem—called evoked potentials— slow slightly.

Allan gets up, looking sleepy. It's noon now. The technician removes the electrodes. He moves to a computer, where he sits on a pillow so he can reach the keyboard. His mission is to navigate using arrows to find a platform in a pool of water. This tests his ability to develop a strategy. He performs it over and over before taking a half-hour break for a pizza lunch. A little over three hours into the testing, and two more to go, Allan looks exhausted. His mother, Winnie Hansen, strokes and kisses his head. "It takes a long time for such a young child," Hansen says with a sigh.

The study actually began before Allan was born, when Hansen and the other mothers gave hair samples to determine how much mercury their babies were being exposed to. Children like Bjørg and Allan— whose mothers ate low or moderate amounts of whale—generally perform well on the tests, beating most of their peers. Allan, whose mother ate whale meat or blubber two or three times a month while pregnant, was in the middle one-third of exposure. Bjørg, whose mother ate it less often—meat less than once a month, and no blubber at all—was in the lowest one-third. These tests can't say anything about the role mercury played in the mental skills of any individual child. But collectively, when the patterns are explored among hundreds of Faroese children, they say a lot. Children with high exposure perform significantly worse in the tests, although their parents wouldn't necessarily notice the subtle effects in their schoolwork or home life. In essence, the findings suggest that a child is a bit younger mentally. The Faroese mothers say they live with the anxiety and guilt of knowing that something they unwittingly ate could have impaired how successful their sons and daughters are in the tests—and perhaps in life.

After lunch, Allan moves to another round of tests of his memory, reaction time and hand-eye motor skills. Shortly after 2:30 P.M., four and a half hours after he arrived, Allan moves to the last session—an extraordinarily thorough physical exam that lasts an hour. The pediatrician, Dr. Ulrike Steuerwald, asks him to toss a ball, jump on one foot, touch his nose with his eyes closed, walk on his toes and heels. She checks his ears and eyes, hits his leg with a hammer, pushes on his spine, pulls on his legs, listens to his heart, weighs and measures him. These are normal children but as any parent or pediatrician knows, there are many variations of "normal." Steuerwald is looking for any pattern that might be associated with mercury. She tests ten general areas, including balance, flexibility, posture, and motor function, and rates each one, comparing them to a total score of 58 on her scale. Allan kicks a soccer ball around the office while his mom talks with the doctor about the results.

The last step is the hardest of all for Allan. A nurse takes his arm and draws a large vial of blood. His mother holds him, rocks him in a chair. He looks pale and tired. The tests are grueling, but they are administered with care by Weihe and his staff. The subjects are treated like children, not like guinea pigs, and the parents become loyal supporters of Weihe's research, returning year after year. It's a never-ending rotation, it seems, because the cohorts are so large, encompassing nearly all children of the islands. Weihe's own daughter took the test when she was seven. She happened to be the right age when the pilot study began on the first cohort. Now a teenager, she vividly remembers the psychological and neurological tests. Weihe even has a photo of her with electrodes on her head.

Why is it important to subject children to such intense tests? Of all the evidence of contaminants endangering the people and animals of the far North, the findings unearthed during this research are the most disturbing. The scientists have repeatedly detected effects on children's brains. Mercury is one of only a few toxic contaminants for which there is persuasive scientific evidence that people are being harmed by doses commonly encountered in their environment. Thirty years of tests on lab animals have confirmed the risks of mercury, showing similar neurological effects as those found in the Faroese children.

"There are many uncertainties," says Grandjean, "and most of them mean that we are probably *underestimating* the true toxicity of methylmercury."

Nevertheless, some think Weihe pushed too strongly for dietary guidelines limiting whale consumption. Some worry it will be the end of whaling. He assures them that that's not his motive. But it's part of his job, he says, to speak out about health threats. He feels extraordinarily loyal to the hundreds of children he has tested over the years, and doesn't want to smear their names or exaggerate the effects he has found. "What I tell mothers is that there is a contaminant in their diet that has a detectable impact on their offspring," he says. "Compared to cigarettes or alcohol, their effect is small. But if they can be removed, they should be." He won't say anything to the media or fellow scientists that he won't say, face-to-face, to the families. Yet he is vilified not only by some of his neighbors, but also by some scientists. The research has faced much criticism and skepticism. He shrugs it off, confident of his findings after nearly twenty years of research. "Time will show" what the truth is, he says.

Some medical experts, citing studies out of the Seychelles Islands in the Indian Ocean, say while the concerns about women eating whale may be justified, there is insufficient evidence to suggest that they are justified for people eating fish. A study of nearly eight hundred children in the Seychelles found no link between their mercury exposure and performance on tests similar to those in the Faroe Islands. Seychelles mothers eat twelve meals of ocean fish a week—so much that their mercury levels were similar to the Faroes women, even though fish is less contaminated than whale. "For now, there is no reason for pregnant women to reduce fish consumption below current levels, which are probably safe," Dr. Constantine Lyketsos, a neuropsychiatrist at Johns Hopkins Hospital, wrote in the journal *Lancet* in 2003. Dr. Gary Myers of the University of Rochester, lead scientist for the Seychelles study, said in a journal report that the results "do not support the hypothesis that there is a neurodevelopment risk . . . resulting solely from ocean fish consumption."

But a national panel of experts and the U.S. Environmental Protection Agency disagree. The EPA, with the support of the National Research Council, decided to rely on the Faroes results when setting U.S. recommendations on how much fish is safe for women of childbearing age to eat. A panel of the National Research Council, in a 2000 report on mercury, said the Faroese studies have been under "extensive scrutiny" and "the weight of evidence of developmental neurotoxic effects . . . is strong." Tests on children in New Zealand, Brazil, Michigan, and upstate New York detected similar effects linked to fish contaminated with mercury or PCBs, or both. Scientists are mystified by the contradictory results in the Faroes and Seychelles, saying they could be explained by discrepancies in how the tests were conducted, social disparities between the island nations, or variations in how the body handles low doses eaten every day compared with high but less-frequent doses. "When the magnitude of an association is subtle, it is not surprising that it is not detected in every cohort studied," the panel's report says.

The youngsters in the Faroese tests are all considered normal, and the effects hardly noticeable, so are the reductions of IQ important to public health? In the United States, the National Research Council concluded, based on the Faroese tests, that exposure to mercury-tainted seafood in the womb could lead to "poorer school performance." The council's report says "deficits of the magnitude reported in these studies are likely to be associated with increases in the number of children who have to struggle to keep up in a normal classroom or who might require remedial classes or special education."

Joseph Jacobson, chair of psychology at Wayne State University, and his wife, Wayne State psychiatry professor Sandra Jacobson, have researched the cognitive effects of contaminants in the Great Lakes for more than twenty years. Their findings are similar to those in the Faroes: Children of mothers who regularly ate PCB-contaminated fish from Lake Michigan have lower IQs and reduced memory skills. Joseph Jacobson warns that the effects on a child's abilities should not be automatically dismissed as trivial. He worries about the impacts on a few of the most highly exposed children. "They might have deficits that are severe enough to markedly affect their day-to-day function," Jacobson says. Dr. David Bellinger, a neurologist at Harvard Medical School who served on the National Research Council's mercury committee, says that such IQ differences can affect not just individuals but entire populations by shifting averages on tests and blurring the "not-so-bright line separating normal and abnormal."

The societal impacts of mercury "can be tremendous," according to a report in the journal *Environmental Health Perspectives* co-authored by Deborah Rice, formerly with the EPA. A universal drop of five IQ points would double the number of U.S. children with IQs under 70 who need remedial education.

Four years after Allan and Bjoerg were born, the Faroese home rule government issued an advisory to all women of childbearing age to stop eating pilot whale.

It was not an easy decision, says Prime Minister Anfinn Kallsberg. "The pilot whale has been the backbone of our culture and very important for our people's diet throughout the centuries," he says. But, he adds, he is "quite convinced" that children are at risk so his government acted to protect them as soon as the first results were published in 1998.

Faroese women have mostly heeded the instructions. "I don't think I would eat whale meat again, not until I'm done having children," says Mariann Poulsen, who ate it once a month while pregnant with her son ten years ago. To keep the whaling tradition alive but still protect her daughter, Gudrig Andorsdottir, a nutritionist in Tórshavn, prepares whale for her son and husband once a month, but not for herself and her teenaged daughter—at least until they reach menopause. "How do you explain to your daughter that you need to learn how to cook and prepare this, but you can't eat it until you are forty-five or fifty?" she says. Her daughter, her second child, "has never tasted pilot whale or blubber. Never." Unlike her brother, she wasn't exposed in the womb because Andorsdottir had stopped eating it by then. It wasn't easy for her to give it up. Andorsdottir came from a village that slaughtered lots of pilot whales, and she ate it once a week as a child. It is a community event and everyone looks forward to the fresh meat, spending days salting and freezing it. As a nutritionist, she knows that the whale contains lots of protein, niacin, iron, and fatty acids. It's one of the most nutritious foods around and, she says, it has been the salvation of her people. When she heard about the high mercury levels in whales in the mid-1980s, she was shocked, and began to eat less, although she still worries about people substituting unhealthy processed foods for whale. "It's quite important to us, as a country, not to lose the knowledge as to how to kill the pilot whale, how to store it and use it. It's been so helpful to us over the years and really our only free food," she says. "Now we are learning while our fish is healthy, our whale may not be as healthy." Whether people follow the advice is "very individual," she says. "Some do, some don't. A lot of people have listened to the results because they are very alarming."

One who doesn't heed them is Jorgen Niclasen, the Faroes' former minister of fisheries, which includes whaling. He eats the blubber weekly and the meat twice a month. His wife ate it while pregnant in 2001, several years after the warnings were issued. "We've been eating this for ages. I believe there are negative things but there are positive things, too," he says. "My wife decided she would take that risk because it's such a good food. If it came to a point we can't take whale, we can't take fish, we would just have to shut down and hope the other countries will take us. The only reason we are here is because of the ocean and what it gives us. We don't live here because of the weather."

Born in a village of seventy people, Ólavur Sjurdarberg is chairman of the Pilot Whalers' Association. A fisherman, school principal, and teacher, he eats whale blubber and meat twice a month, sometimes as often as ten times a month after a slaughter. His daughter and two sons also eat it. Like many Faroese people, he craves the taste and the energy it provides. "Sometimes I feel like I'm hungry for it, like you get hungry for chocolate." As a teacher, Sjurdarberg wonders about the effects on children. "I can't believe it. Or maybe I will not believe it. I don't know," he says. He worries most that the contaminants will remain in the ocean for generations and might sicken whales or kill the krill that supports the abundant fisheries. "And then maybe all the creation in the oceans will get problems," he says. "Even if we stop eating whales, the pollution is still there. You can't tell the whales not to eat it," he says. "We have no garden we can harvest. The ocean is our garden and we have to protect it. We'll never take the last whale. We'll never take the last fish. We'll never take the last bird. We would be killing ourselves. Life here in the Faroes would be finished."

Jacob Pauli Joensen, head of environment for the Faroese Home Rule government, says his agency was "very reluctant" at first to accept the findings and issue a warning. They worried that people around the world would misconstrue it as a warning against eating Faroese fish, when it applies only to whale. "We were not especially happy about" the Faroes being a laboratory for studying children, he says, "but we can't hide. The best thing we could do was to react." In 1998, warning letters were sent to all women between the ages of twenty-five and twenty-nine, the most frequent childbearing years. Since then, the advice has been widely circulated. It is impossible to live in the Faroes and be unaware of it. "We hope this sends signals to the outside world that we are very afraid of this pollution," Joensen says.

The government warnings worked. Mercury levels of Faroese women are one-tenth of what they were in 1986, when the children in the first cohort were born, and they are now on par with the rest of Europe and the United States. Yet excessive mercury exposures are still occurring further north, in the Arctic region. Mothers in Qaanaaq, Greenland, carry the highest mercury concentrations recorded, except in industrial accidents such as the one at Japan's Minamata Bay. Tests of children there showed similar effects as in the Faroes. Children with high mercury levels had slower reaction times and worse performance on hand-eye coordination tests, as well as measurable delays in signals sent from the ears to the brain.

Because emissions are increasing and concentrations rising in some ocean life, mercury is a growing threat in many parts of the world, particularly in the Arctic, where ringed seals and belugas carry up to four times more than they did twenty-five years ago. Some 5,000 to 10,000 tons of mercury enter the atmosphere every year, 50 percent to 75 percent of it from human activities. By far, the biggest man-made sources are coal-burning power plants and waste incinerators. Every year, they release about 1,500 tons worldwide, more than half of it in Asia, the United Nations Environment Programme (UNEP) says in its 2003 Global Mercury Assessment. The United States, in comparison, is a much smaller part of the problem. Its coal-fired power plants, mostly in the Midwest and Northeast, emit about 3 percent of the global emissions, or forty tons a year. The U.S. EPA has proposed regulations that would cut mercury from power plants, but environmental groups say they don't cut deep enough or fast enough. Internationally, there are no treaties in place to reduce global mercury emissions.

The typical American carries ten times less mercury than the mothers of the Faroese children who were the subjects of the tests that detected the reduced intelligence. But in many regions, including California, people who frequently eat ocean fish are exposed to just as much mercury, sometimes more. "The available data indicate that mercury is present all over the globe, especially in fish, in concentrations that adversely affect human beings," says the UNEP's mercury assessment.

Even in the United States, there are extremes in exposure, depending on how much and what fish are eaten. One San Francisco physician, Dr. Jane Hightower, discovered that excessive levels are common in upper-income women and children. One of her patients, a woman who ate swordfish purchased from restaurants or stores fourteen times a month, suffered memory lapses and her hair began to fall out. A toddler who regularly ate salmon and sole had mercury levels over twice the recommended amount, Hightower says. "It's the dose that makes the poison, but what makes mercury a concern is that it is ubiquitous in fish," says John Risher, a mercury specialist at the toxic substances agency of the U.S. Centers for Disease Control and Prevention. "There are a lot of people that eat a fair amount of fish and it can build up over time."

One of every six women in the United States—about 630,000 pregnant women a year—carry more mercury in their blood than the 1 part per million standard that the EPA recommends to protect fetuses. But this doesn't mean that women should stop eating fish and other seafood, which are healthy and nutritious foods. Rather, the U.S. Food and Drug Administration has advised women of childbearing age to avoid the most highly contaminated fish—swordfish, shark, tilefish, and king mackerel—and limit all other fish consumption to twelve ounces per week, typically two small servings. If the U.S. advisory works as well as Weihe's did in the Faroe Islands, then mercury levels in American children should begin to drop.

The prime minister of the Faroe Islands says that while the youngsters of his homeland have been in a unique position to chronicle mercury's hazards because of the testing there, they are merely messengers of a global threat to children all over the world.

"We are victims of the pollution that other nations create," Kallsberg says. "We ourselves do not have these heavy industries that leak these poisons into the ocean. Once it's released, it's not anymore their problem. It becomes ours."

It is a warning that echoes one that Rachel Carson sent nearly a half-century ago, when chemicals were beginning to build up in human bodies.

Chapter 9

Beyond Silent Spring: A Global Assault on Sex Hormones and Immune Systems

For the first time in the history of the world, every human being is now subjected to contact with dangerous chemicals from the moment of conception until death. . . . They have been found in fish in remote mountain lakes, in earthworms burrowing in soil, in the eggs of birds— and in man himself. For these chemicals are now stored in the bodies of the vast majority of human beings, regardless of age. They occur in the mother's milk, and probably in the tissues of the unborn child.

When Rachel Carson wrote those words in 1962 in her best-selling book *Silent Spring,* no one knew the biological ramifications of chemicals found in the tissues of wildlife and in the fat, blood, and breast milk of people. If birds are falling from the sky, dead, what does this mean for the rest of us? Could this explain the epidemic of cancer? Will these chemicals kill us? "Elixirs of death," Carson called them; a "grim specter." Whatever the chemicals were doing, society decided that it couldn't possibly be good for people and nature. So, in the face of such fears, the leaders of the industrialized world acted swiftly. In the 1970s, many nations banned the pesticide DDT, a handful of related organochlorine pesticides, and industrial compounds called PCBs, all chemicals that were amassing in the environment. The problems, however, were far from over. In fact, they were just beginning.

Only in the past decade have scientists realized how insidious the damage inflicted by toxic pollutants really is. The danger goes far beyond cancer, far beyond pesticides. Scientists have now learned that many chemical contaminants are capable of assaulting the innermost workings of living things—their sex hormones, their immune cells, their brains—interfering with processes so vital to life on Earth that Carson never imagined such dangers, or at least never dared to mention them. Some of the chemicals, most notably DDT and PCBs, have endured in the environment longer than anyone imagined in the 1960s. Some have new names—PBDEs, PFOS, phthalates—but still follow the same familiar patterns.

Several dozen chemicals, some still in use today, are known to mimic sex hormones, causing a variety of animals to be born with reproductive oddities. And immune systems are under global assault from chronic buildup of contaminants that can deplete antibodies and exacerbate disease. Scientists used to talk about cancer when they warned about the long-term effects of exposure to toxic chemicals. Today, cancer is often the least of their worries. The most sensitive targets, scientists now say, are babies exposed to contaminants that can alter the brain, as shown in the Faroe Island children, or disrupt sex hormones and immune cells.

After World War II, chemists began manufacturing chemicals that, in the words of Rachel Carson, "nature never invented." New synthetic compounds were being created on a daily basis, especially after the wonders of combining carbon and chlorine molecules were discovered. Production of insecticides soared in the 1950s. DDT, chlordane, heptachlor, dieldrin, toxaphene, mirex, endrin, aldrin—they all were cousins, all organochlorines used as pesticides. PCBs, another chlorinated product, were marketed as perfect fireproof insulators and hydraulic fluids. The power of the chemicals, particularly the pesticides, was thought to be extraordinary. They weren't immediately or acutely poisonous, so they seemed relatively benign. Scientists know how chemicals react in a laboratory, but they often are astonished to learn how they act in the environment. They used to think chlorinated chemicals such as DDT and PCBs were safe to marine ecosystems because they were not acutely poisonous. They were blinded by the chemicals' inertness. Soon, though, it became clear that they were dangerous in a slow, insidious sort of way. They didn't easily break down in the environment. They were collecting in bird eggs, in animal fat, in human tissues. They were being passed from one animal to another: from plankton to worm to fish to bird, from hay to cow to milk to child. Birds suffered the brunt of it. Some dropped dead. Some hatched eggs with shells too thin for the chicks to survive. Some were born with severe defects like crossed bills. Even the ubiquitous robin and the mighty bald eagle were on the verge of extinction. By the mid- 1970s, many of the chemicals had been banned, and experts thought the problem was solved.

The first hint of deeper trouble surfaced off the coast of Los Angeles. Seagulls on Santa Barbara Island—a major breeding area—were showing peculiar sexual behavior. Females were pairing up and sharing nests. Upon closer examination, two Southern California scientists found the reason. They counted as many as nineteen female gulls on the island for every male. Where, they wondered, had all the males gone? Because of the skewed gender ratio, the island's colony of Western gulls started to collapse, dropping from three thousand pairs in the 1950s to about seven hundred by the mid-1970s.

Avian toxicologist Michael Fry of the University of California at Davis had a theory. He wondered if DDT, notorious for thinning the eggshells and killing the embryos of Southern California pelicans and cormorants, could be to blame for the missing males. A ton of DDT was still lying off the coast of Los Angeles, discharged there by pesticide manufacturer Montrose Chemical Co. Gulls, unlike pelicans and cormorants, had appeared to be unharmed by DDT; they had no visible birth defects or other physical disorders. But Fry was shocked to find, upon close inspection, that many gulls had both testes and ovaries. To confirm his finding, he injected gull eggs from an uncontaminated lake with amounts of DDT comparable to that found on Santa

Barbara Island. In his laboratory, he had crafted hermaphroditic birds. DDT had neutered them. Fry's work—published in the journal *Science* in 1981—was the first evidence that environmental contaminants were feminizing wildlife.

Most scientists shrugged it off as an inexplicable quirk in a uniquely contaminated spot. At the time, common wisdom held that contamination that did not kill or maim animals was harmless. If an animal was alive, it was assumed to be healthy and fertile. With DDT and PCBs banned, scientists lost interest and moved on to studying other environmental issues.

About a decade later, in the late 1980s, zoologist Theo Colborn of the World Wildlife Fund was researching a book on Great Lakes wildlife when she began turning up studies indicating that more than a dozen birds, fish, and mammals were suffering reproductive ailments and birth defects linked to heavy DDT and PCB contamination. The oddest part, she thought, was that the problems weren't showing up in the adults that had grown up in the 1960s and 1970s, when pollution had peaked. They seemed fine. Instead, cormorants born more than a decade later had twisted, mutated bills, and the only healthy eaglets seemed to have parents that had just migrated to the lakes. "It was very evident that the problems affecting wildlife around the Great Lakes were not in the adult animals, but their offspring. The offspring that survived were not in very good shape," says Colborn. Tapping into an electronic database, Colborn pulled together hundreds of scientific reports. She learned that reproductive problems had been detected in almost every developed country since synthetic chemicals were introduced after World War II. But because the reports came from scientists from a wide array of disciplines who rarely see or hear about one another's work, no one had noticed the similarities. "By the fall of 1988, it became very overwhelming that there was something really wrong," Colborn says. "As I put it all together, I had to do something with it." Finally, in the summer of 1991, Colborn convened a meeting of twenty-one North American and European zoologists, pathologists, immunologists, and other scientists researching the endocrine systems of humans and animals. After three days, they issued an alarming six-page statement. "Many wildlife populations are already affected by these compounds," the scientists concluded. "It is urgent to move reproductive effects to the forefront. . . . A major research initiative on humans must be undertaken."

They had reached the frightening conclusion that chemicals were capable of imitating hormones, a phenomenon that would soon be dubbed "endocrine disruption."

The hormone problem sprang to life a few years later, in the gender-bending waters of Lake Apopka, on the suburban outskirts of Orlando. Florida scientists, led by Louis Guillette and Tim Gross, discovered that alligators in the lake weren't quite male but they weren't quite female either. They were both. Or neither. "Everything is really fouled up. It is indeed real, and it is scary," says Gross, a University of Florida wildlife endocrinologist. "We didn't want to believe it, in all honesty."

This is no fluke of Mother Nature, no quirk of evolution. It appears to be a legacy of pollution. In 1980, Tower Chemical Co. spilled a large quantity of kelthane, a pesticide containing DDT, along Lake Apopka's banks. Fertilizers and insecticides routinely flow off nearby citrus groves and other farms, while little fresh water is allowed in. Even Hurricane Andrew took a swipe at Apopka and stirred up its polluted sediments. "Everything that could happen to a lake has happened to this one," says University of Florida alligator biologist Ken Rice.

Since the early 1980s, alligator births at the lake had dropped five-fold, which prompted the scientific team to begin exploring the cause. The first breakthrough came in the fall of 1992, when Gross, conducting routine sex typing on Lake Apopka's turtles, couldn't tell males from females. The following March, Guillette and Gross examined six-month-old alligators hatched in their laboratory from eggs taken from the lake. Males had the high estrogen and low testosterone typical of a female, while females had double the normal estrogen and were laying clutches that, instead of containing the usual one egg, had as many as seven, all dead. A few months later, the scientists confirmed sexual abnormalities in alligators born in the wild—four of twelve had part-male, part-female genitals. "It sent up signals and flags and fireworks and we knew something was going on there," says Guillette, director of the University of Florida's reproductive analysis laboratory. "We now know that every single animal that we thought was a female in Lake Apopka had abnormal gonads."

DDE, a metabolite of DDT, was found in the tissues, and the team replicated the link in the laboratory when they painted alligator eggs with minute amounts—as low as 10 parts per billion—and the young were born with abnormal hormones. In comparison, not a single alligator at nearby Lake Woodruff, an uncontaminated lake in a national wildlife refuge, had hormonal or sexual defects. Still unconvinced, the researchers returned to both lakes. At night, they kicked around every imaginable theory, even an unlikely one they jokingly named the "shanty-town girl hypothesis"—that perhaps Lake Apopka is a magnet for weak female alligators. "We wanted to be 100 percent sure," Gross said, "before we opened our big mouths."

In August 1994, they finally published their findings in a scientific journal, warning that their two years of studies "have only begun to address what we believe could be a serious, widespread threat to wildlife populations" and perhaps humans. Guillette was vocal and omnipresent, vocalizing his concerns in a straight-shooting way that prompted headlines in the mid-1990s and captured society's attention. The potential consequences, he warned, are almost unthinkable. If males aren't male and females aren't female, they cannot reproduce, and some outwardly healthy populations could be a generation away from extinction.

Elsewhere around the world, the same phenomenon of sexual confusion in the wild began turning up in a menagerie of fish, birds, and other animals—eagles in the Great Lakes, carp in Nevada's Lake Mead, river otters and salmon in the Pacific Northwest, fish in Great Britain—and polar bears in Svalbard, near the North Pole. Testosterone levels plummeted in some male animals, while females were supercharged with estrogen. Both sexes sometimes are born with both a partial penis and ovaries, and some males, particularly fish, wind up so gender-warped that they try to produce eggs. Pesticides and industrial chemicals can infiltrate wombs and eggs, where they send false signals imitating or blocking the hormones that control sexuality. Estrogen and testosterone are the body's sexual messengers, ordering embryos how to grow. When a pregnant animal is exposed to even a minuscule dose of a hormone—real or fake—during the onset of its embryo's sexual development, the gender of the offspring can be irreversibly altered. In their 1996 book *Our Stolen Future,* Colborn and coauthors Dianne Dumanoski and John Peterson Myers call them "hand-me-down poisons." Although the parents are unharmed, their offspring's development is disrupted.

No one knows what this portends for humans—who often encounter the same chemical residues in their food and water. Because hormones play a role in the human body identical to that in other creatures, some scientists are alarmed by the potential of adults passing fertility problems and other reproductive defects on to their children. The ability of these chemicals to leave the parents unharmed but afflict the unborn is one of the most critical and hotly debated environmental issues of modern times. Used in pesticides, plastics, detergents, and industrial compounds considered critical to a variety of activities, endocrine disrupters are ubiquitous. The human toll from hormone-altering chemicals in the environment remains shrouded in scientific doubt and controversy, and credible answers could take generations of research. Scientists agree that pesticides and other pollutants are feminizing fish, birds, and other creatures, but they have reached no conclusion on whether the hormone-mimicking pollutants are harming people too. There are so little data and so many lingering questions that it is impossible to tell. Many biologists theorize that pesticides and industrial chemicals might be lowering sperm counts, triggering diseases in ovaries, breasts, prostates, and testes, raising the risk of osteoporosis, retarding penis growth, and causing children to reach puberty prematurely. But such links are highly speculative and fiercely debated.

The most heated argument is focusing on whether sperm counts are declining. In 1992, reproductive biologists in Denmark thought they had an answer. After examining medical reports of almost 15,000 healthy, fertile men from many nations between 1938 and 1990, they found sperm concentrations were almost half what they had been— 66 million per milliliter in 1990 compared with 113 million per milliliter in 1938. The findings spurred concern that post–World War II chemicals and pesticides were to blame. Researchers already had been worried after small studies showed that some industrial and farm workers handling large quantities experienced sperm problems. Others, however, point out that the steepest drop in sperm occurred in the 1940s and 1950s, before contamination peaked, and that no declines have been found in some regions of the world with heavy pollution. Even if sperm counts are decreasing, proving the cause is virtually impossible because men are exposed to myriad factors that might be to blame, from hot tubs to venereal disease.

If not for the real-life saga of DES mothers, it might seem far-fetched to believe humans could pass chemical-induced sexual disorders or fertility problems on to their sons and daughters in the womb. DES, or dicthylstilbestrol, was a powerful synthetic estrogen prescribed to 3 million to 4 million pregnant women between 1948 and 1971 to prevent miscarriages. Although the mothers were unscathed, their daughters suffered a high incidence of ovarian abnormalities and vaginal cancer and their sons have abnormal rates of infertility and testicular disorders. Still, most estrogenic pesticides and industrial chemicals are hundreds of times less potent than DES, and people also routinely consume some natural hormonal chemicals found in plants with no apparent harm. The big unanswered question is whether an average person's diet, which is much more varied than an animal's, contains large enough quantities of man-made hormones to endanger an unborn child.

Some of the most disturbing findings related to hormone disruption are in the Arctic. Andrew Derocher and his colleagues found altered testosterone and progesterone in Svalbard's polar bears. The bears are able to reproduce—many of them have cubs—but scientists worry that the hormone changes could be altering their fertility and leaving bear populations smaller than expected there. No research, however, has been done, at least not yet, on the human population of the Arctic to check whether their hormones are normal and their fertility intact.

Sex hormones aren't the only body function seemingly under attack from endocrine-disrupting chemicals. They also tamper with thyroid hormones that regulate neurological development of a fetus, much like mercury does. But it is the assault on the immune system that poses one of the most worrisome threats to human health and wildlife. As with polar bears and sex hormones, it took the ocean's predators—seals and killer whales—to provide scientists a living laboratory to document that the body's protection against disease is weakened by PCBs.

In 1988, death came abruptly to plump harbor seals basking on the tiny Danish isle of Anholt in the North Sea. First the pups, then the adults died. Their lungs clogged with fluid, skin became mottled with ulcerous sores, and fevers soared. They could barely swim and refused to eat. Within days, almost 300 carcasses piled up along the shore of this sparsely populated retreat, the most isolated island in Denmark, where seals breed every summer. With the virulence of a hurricane, the mysterious killer jumped south, then north, then west, until it had ambushed nearly every seal colony of northern Europe. When the outbreak ended six months later, 20,000 seals had perished—more than half the continent's population. Europeans were horrified. It was an ecological disaster of epic proportions—orders of magnitude worse than any oil spill—and they wondered what had gone dreadfully wrong along their cherished coasts. Speculation raged: Poisonous algae. Global warming. A chemical spill. When, months later, the culprit was identified as a newly discovered distemper virus, Europeans were relieved that the deaths could be chalked up to "natural causes."

Peter Ross, though, questioned whether it was, in reality, natural. He was studying harbor seals in Nova Scotia at the time of Europe's plague. Why, Ross wondered, were the seals in eastern Canada healthy and thriving? Tests of their antibodies showed that they had been exposed to the same distemper virus but, for some reason, the Canadian seals fended off the disease. Ross and other marine biologists suspected that something—most likely PCBs, which were at high levels in the North Sea—was suppressing the seals' immunity. But they had no proof, except for some tests on lab rats that showed PCBs depleted immune cells. What they needed were measurements of seals' immunity under real-life conditions. Such an experiment on wild animals had never been conducted. Ross and his colleagues decided to try.

Three years later, under the auspices of the Dutch government, Ross moved to the Netherlands and embarked on a project to raise seals in a captive environment. Seal pups were caught off Scotland and divided into two groups kept in separate pens. One group was fed herring from the heavily contaminated Baltic Sea, while the other ate herring from the cleaner Atlantic. The difference in PCB intake was tenfold. After two years, Ross and his colleagues compared the two groups' immune cells. The differences, reported in 1995, were far greater than anyone had predicted. Seals fed the Baltic fish produced 25 percent fewer "natural killer" cells—the first line of defense against viruses—and 35 percent fewer T cells, the white blood cells essential to clearing infections and ordering production of antibodies. Such a severe loss of immunity is comparable to what is seen in some AIDS patients. The scientists, for ethical reasons, did not take the logical next step and expose their subjects to disease. The seals, Ross suspects, all would have died.

Like soldiers on the front line, immune cells defend the body against a foreign invader such as a virus. The immune system, like any good army, has multiple layers of defense. "Natural killer" cells are powerful, fast-moving warriors that mount the first attack against viruses and tumors. T cells clear an infection and order B cells to unleash antibodies, the ammunition against specific foreign agents. But chemicals like PCBs can block the cells from proliferating and mobilizing, disarming this immune infantry. Animals with compromised immunity are more likely to become ill and die when exposed to disease. Without the PCBs in their bodies, scientists believe that northern Europe's harbor seals may have been able to shrug off the virus. What turned into a killer plague might have been no more deadly than the common cold. "You don't get that magnitude of a die-off—twenty thousand seals in Europe—without a contributing factor from the environment," says David Ferrick, a University of California at Davis immunologist who studied California sea otters, dolphins, and seals. "They seemed healthy, but when they were introduced to [a virus], it just decimated them. That is condemning evidence that immune impairment is one of the few rational explanations."

In wild seals, the immune damage is probably even worse than what Ross found in captive seals. The captive ones were fed Baltic herring for only two years, while wild seals live thirty to forty years, amassing ten times higher levels of PCBs than the 17 parts per million in the captive seals. In comparison, PCB levels in Arctic polar bears have reached as much as five times higher than the levels in the captive seals. "There is very strong evidence that animals in the wild are immunosuppressed. The harbor seal is the canary in the coal mine," Ross says. "When 60 percent of a population dies, it means there is something wrong." The most disturbing implication reaches even further. The fish eaten by the captive seals had been purchased in the Netherland's commercial markets. It is the same Baltic Sea herring served at the dinner tables of European households. "Some people say, 'Who cares about cute, cuddly seals?' But what it indicates is the state of the world's oceans, and the commercial fisheries too," Ross says. Seal die-offs on Anholt Island and the surrounding area are continuing, with tens of thousands more dying in 1998 and 2002 epidemics.

Soon after the landmark study that showed PCBs suppressed the immunity of the seals, Ross moved to the Canadian government's Institute of Ocean Sciences on Vancouver Island. In 1996, he opened a long-forgotten file of another scientist and scanned columns of data inside. A number caught his eye: 250 parts per million. At first, he didn't believe it. These were the highest concentrations of PCBs he had ever seen, many times higher than the amount that suppressed the immunity of seals. They had come from live killer whales, the black-and-white icons of the Pacific Northwest and British Columbia that were swimming in the scenic waters off Vancouver Island. "My jaw dropped," Ross says. "I said, these animals are really hot."

He knew the only animals known to contain more PCBs than these killer whales were already dead—Mediterranean dolphins that died en masse from an epidemic. He was struck with a sudden fear: Could the Pacific Northwest's killer whales be a mere virus away from a mass die-off like Europe's seals? Ross decided to investigate. He pulled the blubber biopsies of thirty more whales from a laboratory freezer and tested them for PCBs. All of the readings were excessively high.

Although feeding in Pacific waters hundreds of miles south of the Arctic, these killer whales are falling victim to the same immune-altering phenomenon as the polar bears and the Inuit of the far North. The orca is a master predator. It glides like a torpedo, its six-foot dorsal fin slicing through the surface of the sea. It hunts down a seal, rams it repeatedly with its tail, and drowns it. These wolves of the sea stand unrivaled at the top of the food web. But like polar bears, their rank in the ocean's hierarchy has given them a perilous distinction. Concentrations of industrial chemicals in killer whales are the highest found in any living mammal, even higher than in Svalbard's polar bears. Male killer whales in the Pacific Northwest contain as much as fifteen times more contamination than Ross's captive seals that suffered the weakened immunity. At those concentrations, the whales "greatly exceeded many toxic thresholds for mammals," Ross says. Killer whales, like polar bears, carry such extraordinary loads of chemicals because the animals they prey on are high on the food web and because they consume so much food—200 pounds a day. Also, the longer an animal lives, the more contamination it stores. Some of the region's killer whales were born before World War I. "They are like sponges that essentially soak these chemicals up," Ross says. "It has nothing to do with how close you are to the pollution source, but how high you are on the food chain."

Two types of pods frequent the waters around Washington's San Juan Islands. The "transients," which prey mostly on seals and sea lions, are the most highly contaminated. The "residents" eat only fish—mostly chinook salmon. The killer whales apparently are getting the chemicals from salmon and seals. But where are the salmon and seals picking them up? Some clearly come from local waters, particularly Seattle's Puget Sound, where industrial and port operations deposited PCBs before they were banned. Asia also is probably contributing quite a bit to the problem. Airborne contaminants blow from China to North America in about a week. Salmon also pick them up from Asia when they migrate into the North Pacific.

Long-lived, elusive, and intelligent, these whales have no predators. Nothing at sea is capable of killing a killer whale, except a human being. But the region's killer whales have been dying at a higher rate in recent years, most disappearing without a trace. Nearly half of their calves die within months of their births. On a typical summer day, hundreds of tourists and boaters set sail in hopes of spotting killer whales. Lately, though, there have been more whale-watching vessels than whales plying the picturesque waters between Seattle and Victoria. The region's famed orca pods are shrinking, with only eighty-four whales left in 2004. These descendants of Sea World's Shamu, revered in native mythology as supernatural in their survival skills, will soon be officially listed as an endangered species. What is killing the resident whales is unproven, and there are no signs of a die-off. "Usually the animals just disappear. We rarely recover carcasses," says Graeme Ellis, a researcher at Canada's Department of Fisheries and Oceans who has been studying the region's population since the 1970s. Some scientists speculate that immune suppression from PCBs has something to do with the orcas' decline, but as with Norway's polar bears, proof eludes them. Ross and colleagues reported in 2004 that the PCBs "represent a tangible threat to (the orcas') health, and may lead to diminished reproductive performance, neurological deficits and increased susceptibility to disease."

Worldwide, viral epidemics among wild animals are spreading farther and faster than ever, especially among dolphins, seals, birds, and other fish-eating animals that carry high body burdens of immune-damaging chemicals. The distemper outbreak that decimated Europe's seals actually began a year earlier thousands of miles away, first striking Siberia, then the east coast of the United States. A record number of dolphins washed ashore between New Jersey and Florida in 1987. As much as half of the near-shore population of bottlenose dolphins was wiped out. Three years later, the distemper plague struck the scenic beaches of Spain, France, and Greece, with more than one thousand striped dolphins piling up on the shores of the Mediterranean. In 1994, along the Gulf shore of Texas, bottlenose dolphins again died in record numbers. Scientists found that the dead animals shared a common bond beyond the virus—their bodies carried high amounts of PCBs.

Scientists now suspect that "natural causes," the traditional and simplistic explanation for the mass epidemics, masks the underlying man-made cause, immune-altering pollution. One of the most sensitive parts of the body—whether seal, fish, bird, or human—is the immune system. Minute levels of contamination—too small to poison, maim, or mutate—seem to suppress the volumes and efficiency of animals' immune cells. Only recently have scientists developed tools— many borrowed from AIDS research—sophisticated enough to go beyond simple counts of white blood cells. The detectives in this rapidly growing field, called immunotoxicology, now can measure the exact and subtle changes in individual immune cells and functions. Using these techniques, they have found that polar bears in Svalbard have fewer immune cells and antibodies if their bodies contain high levels of PCBs. But a key question remains: Are these chemicals disarming natural defenses so severely that disease turns deadly? For wild animals, most experts believe that it is a question of degree—not whether pollution is weakening their immunity and contributing to their deaths, but by how much. The prevailing wisdom is that many exposed creatures have become the animal kingdom's version of AIDS patients—they die from infections they could have fended off if their immune system hadn't been compromised. "Are certain pollutants at certain levels causing effects on the immune system? Absolutely," says Judith Zelikoff, a New York University Medical Center immuno-toxicologist. "We know they are bringing about changes. And there is a strong suspicion, and strong evidence, suggesting that it plays a major role in increased incidence of wildlife morbidity and mortality."

Although sea mammals are most susceptible because they feed their entire lives in polluted oceans or estuaries, humans could be at risk too. The immune system of a human being is virtually identical to those of other mammals, and they encounter the same immune-suppressing chemicals in their food. In the Seattle area, for example, people eat the same salmon, albeit in much smaller quantities, as killer whales. "What's going to hurt a marine mammal," Ferrick says, "is probably going to hurt us too."

In the Arctic, people routinely encounter high levels of industrial compounds and pesticides that deplete the immune cells of marine mammals and laboratory animals at fairly small doses. For most healthy people, a slight drop in immunity caused by the pollutants carried in their bodies merely could mean they catch the flu more often or stay sick a bit longer. But for vulnerable newborns or the chronically ill—especially those with the AIDS virus or other immune deficiencies—it could seriously compromise their health, immune experts say. Suspicions about the human effects are unproven, yet the theories are bolstered by a growing body of evidence from researchers, particularly in the Arctic, where children suffer inordinate amounts of respiratory and ear infections. Scientists are now exploring whether Arctic children have reduced antibodies, a telltale sign of immune suppression.

For wildlife at least, death due to "natural causes" often has more to do with the acts of man than the acts of nature. "Disease is an expression of an environment out of balance," says Milton Friend, director of the National Wildlife Health Center, a U.S. laboratory that investigates animal epidemics. "And nothing is natural out there anymore."

While biologists continue to probe bodies for the insidious health effects of global contaminants, chemists and physicists are trying to unravel how the Arctic has become the world's wastebasket, the last refuge for toxic garbage that originates thousands of miles away.

Chapter 10

The Arctic in Flux:
Global Conspirators and
the Whims of Climate

On a late February morning in Canada's High Arctic, the orange glint of sunrise creeps over the horizon, setting the dark, indigo sky aglow. A yellow streak of sunlight blazes across the sky, then vanishes. The day ends—the period from dawn to dusk spans only an hour or so—and twilight descends, all too soon. It is the first day of polar sunrise, a brief reawakening after the sun's months-long winter absence. Over the next two months, the sun will stay a bit longer each day, until, in March, the days are longer than the nights. Then, in October, darkness descends again, enduring for almost five months.

Strange and baffling things happen during these prolonged periods of darkness and light. Chemicals transform themselves in unexpected ways, moving, reacting, morphing—even vanishing *from* thin air.

In 1995, scientists from Environment Canada discovered a phenomenal vanishing act at Alert, a weather and military station at the northern tip of Nunavut's Ellesmere Island, just 500 miles from the North Pole. The team, led by meteorologist William Schroeder, was using state-of-the-art, automated atomic fluorescence monitoring devices that allowed them to take continuous air samples every half hour. Throughout the long winter, the graphs had shown stable readings for mercury vapors, hovering right around 1.5 nanograms per cubic meter of air. Suddenly, in March, the graphs began fibrillating like the EKG of a heart attack victim. Within a matter of hours, the mercury concentrations fell to zero—there was no mercury in the air at all. None.

The chemists were astonished. They knew that it was impossible for mercury to disappear without a trace. Mercury is an element, which means that it cannot be created or destroyed; it can only change forms and move around. So if it was no longer measured in the air at Alert, it had to turn up somewhere else eventually. The scientists began searching and found some of it—underneath their feet. The mercury had fallen from the air onto the ground, depositing on the ice and snow—right where polar animals feed and breed. What would cause such a rapid flux? The scientists, upon returning the next year and each year since, realized that the so-called "mercury depletion events" were occurring in Alert at the same time every year, in spring, after the sun reappears.

For scientists, it was a startling discovery. Here comes the sun, there goes the mercury. It turns out that a unique combination of sun, salt, and mercury spurs a photochemical reaction every polar sunrise—a phenomenon dubbed "mercury sunrise." During the winter, mercury, in its gaseous form, flows to the Arctic, carried by prevailing winds from power plants and industries in Europe, North America and Asia. Because it has a long atmospheric lifetime, up to twenty-four months, it can be transported thousands of miles. An estimated fifty to three hundred tons of mercury flow to the far North every winter. There it stays in the atmosphere—until the sun returns. Dosed with solar rays, bromine in the Arctic Ocean's sea salt interacts with ozone, forming bromine monoxide (BrO). Normally, mercury, in its elemental form, stays in the air, in a form that is harmless to people and animals. But the BrO reacts with the mercury vapors, oxidizing it and making it more reactive, what chemists call "sticky." Some drops to the ground and some is carried by airborne particles, changed into a form that is no longer inert and can be taken up by living plants and animals. The amount of mercury in the snow and ice instantaneously increases as much as twentyfold at polar sunrise. In summer, some of it becomes gaseous again and reappears in the air but some that falls to the ground in the spring may not rise up again. Scientists suspect that much of the missing mercury is taken up by marine life.

Schroeder's Toronto team published its findings in 1998 in the journal *Nature*. "People told us we were crazy, that our instruments were broken," says Alexandra Steffen, a research chemist at the Meteorological Service of Canada who worked on the Schroeder team. Since then, they and other scientists, braving the subzero temperatures of polar sunrise, have documented the mercury phenomenon throughout the Arctic, in Barrow, Svalbard, Greenland, and Hudson Bay, and in the Antarctic, as well.

"At first we had a hard time convincing people that this was indeed happening, given the long atmospheric residence time of mercury. We spent three spring seasons convincing ourselves, and then finally others, that this was not an artifact of sampling but a real phenomenon," says Steffen, who has now studied different aspects of the mercury sunrise for a decade. "Our work at Alert has sparked a new field of research science—atmospheric mercury measurements in the Arctic."

Most chemists had no idea that an element could change so rapidly in the environment. "You have a heavy metal flying around in the air, and you believe it is quite inert and then you suddenly have something, within a few days, disappearing. That's amazing," says Michael Goodsite, an assistant professor of environmental chemistry at the University of Southern Denmark. Polar sunrise changes in other contaminants, particularly ground-level ozone, have been demonstrated as well. "A lot of special chemical reactions are going on just during the polar sunrise, triggered by the sun's photochemistry," says Henrik Skov of Denmark's National Environmental Research Institute.

With discovery of the mercury sunrise, scientists think they are beginning to understand why the Arctic's environmental levels of mercury are so high and have been increasing with latitude and over time. As the mercury constantly enters and reenters the environment in cycles of winter and spring, it could explain why concentrations in some Arctic wildlife, particularly Canada's beluga, have been increasing dramatically over the past couple of decades even though mercury emissions from North American industry have been declining for two decades. It appears that the mercury emitted never really goes away. It just keeps cycling through the environment over and over. "PCBs are a thing of the past," Donald Mackay says. "But mercury you cannot ban. Every time you open up the Earth's crust you release mercury."

The polar sunrise may be the reason that mercury levels have not dropped in some parts of the Arctic. "The connection is speculative," says Rob Macdonald of Canada's Institute of Ocean Sciences, "but we do know that there are animal populations in the Arctic that seem to be vulnerable to mercury. They seem to have high concentrations and they seem to be growing and we don't know why."

As the ice melts, scientists suspect that a burst of mercury flows into the upper ocean and slips under the ice, right at the most vulnerable time and place for the Arctic ecosystem. It may accumulate at the edge between ice and open water, where many Arctic animals feed, and come right at the time when the animals are waking up from the long winter and beginning a cycle of breeding and rapid growth. Steve Lindberg of the Oak Ridge National Laboratory in Tennessee and Goodsite reported that the oxidized mercury builds up rapidly in the snowpack in a form that is bioavailable to bacteria, and then "is released with snowmelt during the summer emergence of the Arctic ecosystem." The theory is that the mercury moves up the food web, from plankton to fish to sea mammals to people, increasing in magnitude at each step. "Although poorly understood, this [sunrise] process may be the chief mechanism for transferring atmospheric mercury to Arctic food webs," according to a 2002 AMAP report.

Mercury is a neurotoxin, one of the most dangerous contaminants in the Arctic, so the reports that it falls rapidly to the ground like a spring shower and is deposited in a bioavailable form in the food web every polar spring are disconcerting. International teams of scientists are now investigating what effects this cycling might have on the reproduction and health of marine life. "One thing is definite," Macdonald says. "The Arctic is an especially sensitive location on the globe to mercury."

The sunrise phenomenon suggests that the Arctic is important to how mercury moves worldwide. "The enhanced deposition may mean that the Arctic plays a previously unrecognized role as an important sink in the global mercury cycle," according to the AMAP report. Mercury doesn't leave a fingerprint, whether it's from a power plant or a mine or a natural source. As a result, it is difficult to trace. No one knows, for example, where the mercury that gets in the pilot whales consumed by people in the Faroe Islands or the beluga in Nunavik or the narwhal in Greenland originated. Before the industrial age, mercury was all locked up in the ground, but when coal is burned, the mercury inside it is freed up and released into the air. Asia's power plants are responsible for 860 tons of mercury emissions per year, more than half of the 1,500 tons emitted worldwide and about twenty times more than those in the United States, according to a 2003 UNEP report.

Yet the increase in mercury within the Arctic may have more to do with changing ice conditions and the polar sunrise phenomenon than output by power plants and factories. Lindberg suspects that the mercury sunrise may be a recent phenomenon, driven by global warming. As the ice cover over the Arctic Ocean thins, leaving more open water, more bromine in sea salts could be escaping and reacting with mercury, prompting it to drop into the ocean. Evidence already exists that the world's oceans are getting saltier as they warm. That means mercury levels in the Arctic could keep rising along with its temperature—one of many climactic surprises in store for the far North as the Arctic heats up.

While mercury has presented one of the most interesting discoveries about how and why chemicals cycle globally, every chemical follows its own path. When a pesticide is sprayed on a garden in London or on a farm field in Iowa, where does it go? Chemists say there are a variety of "choices," dictated by the laws of nature. All molecules move at different rates, accidentally, in a random way, "like kids when they arrive at Disneyland," Mackay says. Some will move rapidly toward their ultimate destination and others will drift slowly.

The routes taken by contaminants are governed by two main forces that act like "conspirators"—the atmosphere and the oceans. The Arctic's atmosphere is an immense circulating mass of air. By and large, it stays at high latitudes but wind feeds in and out of it "like a merry-go-round with kids coming in on one side and off at another," Mackay says. The pathways the winds take depend on the season. In winter, they tend to move from temperate zones to the Arctic. Every winter, when the Earth's oceans are much warmer than its continents, a large cell of low pressure, called the Icelandic Low, hovers in the atmosphere over the Atlantic. At the same time, a high-pressure system lingers over the continents. This confluence produces powerful westerly winds over the North Atlantic, which blows contaminants from the east coast of the United States and Canada across the Atlantic, toward Arctic Canada, then onto Greenland and continuing east toward Norway. It also generates southerly winds over the Norwegian Sea, which sweeps mainland Europe's chemicals to its High Arctic, where polar bears are born on the archipelago of Svalbard. In a matter of days or weeks, chemicals that originated in the cities of North America and Europe are contaminating the Arctic's air. When they reach the cold air, they condense and drop into the ocean or onto the frozen ground, where they are absorbed by plants, then animals. The colder the ocean and the air, the slower the natural process of decomposition, so most of the chemicals in the Arctic remain there.

Environment Canada atmospheric expert Terry Bidleman says that each contaminant reacts differently in the atmosphere and water. Some can hop long distances; others are "sticky" and fall to the ground. "It's not just a great big soup bowl up there," Bidleman says. Acids and particulates are "single-hop" chemicals, which transfer from soil or water to air only once, while organochlorines and mercury—the most prevalent contaminants in the Arctic—are multi-hop ones. The pesticide hexachlorocyclohexane (HCH) is especially prone to Arctic journeys. Half of all the HCH left in the world is now in polar environments, even though none was ever used there. In 1992, scientists at Japan's Ehime University reported a strange phenomenon: Although HCH levels are highest in the atmosphere in the tropics, where they are mostly used, levels in the water are highest in the Arctic. Logic seems to dictate that HCH and other chemicals would be highest at their source and then decline with distance, yet HCH and other mobile contaminants defy this logic. They increase with distance, which scientists call an "inverted profile." The pesticide is held in the ocean, sandwiched between the surface of the ice and a barrier that exists at a depth of about five hundred feet because of stratification of the ocean. Freshwater is denser than salt water so it sits on top, preventing the downward flow of HCH. Even though all but one form of HCH has been banned in North America, it will keep flowing for years into the archipelago, the Arctic islands off Canada, because of the ocean's propensity to store it. Other chemicals with similar volatility, such as toxaphene, will react in much the same way, continuing to build up.

Surprisingly little is known about where chemicals wind up. Most don't leave a "fingerprint," so it's difficult to trace them back to their sources. Someday, using what they call "mass balance models," scientists eventually hope to track chemicals from cradle to grave—predicting where a chemical goes, how long it will stay, what fraction reaches the Arctic, how much will get into wildlife and humans, and how long before nature eliminates it. Scientists suspect that only a small fraction of the chemicals used in the Northern Hemisphere actually reach the Arctic. The rest spreads all over the world, with some even crossing the equator. "Contrary to common misinterpretation, most of the global chemical inventory of POPs [persistent organic pollutants] will not eventually reach the Earth's polar regions. . . . In reality, most POPs will fall victim to retention and degradation in source areas and en route, and never reach polar regions," chemists Mackay and Frank Wania reported in a 1996 report in the journal *Environmental Science and Technology*. Nevertheless, they say, high concentrations occur in Arctic life "if even a minor share of the global inventory migrates to the polar regions." Imagine, says Mackay, how polluted polar bears and Arctic people would be if every drop used around the world wound up there. Thankfully, he says, that doesn't happen.

The biggest pulse of PCBs and other persistent organic chemicals entered the global environment, mostly from the east coast of the United States and western Europe, in the 1950s and 1960s. Since their ban, PCB emissions have declined dramatically and, consequently, most environmental levels around the world have dropped. But about twenty years ago, they apparently hit a plateau, with today's levels remaining similar in most areas to what they were in the mid-1980s. Environmental samples don't match the downward trends in reportable, new emissions, apparently because PCBs are still being emitted from old electrical equipment and old chemical stockpiles. Scientists have tried to inventory the volumes of organochlorines used globally in the past as well as in the present and track their whereabouts, but it is no easy undertaking. Many developing nations, particularly in Asia and Africa, have hundreds of thousands of tons of old pesticides stored in unsafe conditions. Industrialized nations are continually emitting the old chemicals, too. In the United States, as recently as 1998, industries reported that they released 3 million pounds of PCBs into the air and water. Scientists have found only 2 to 10 percent of the tons of PCBs known to have been produced globally. What happened to the rest? Has nature degraded it, rendering it harmless? Or is it still cycling around the Earth, perpetually in motion, carried by winds and currents and animals? Migrating fish, whales, and birds carry compounds inside their bodies and spread them during their migrations, in a phenomenon called *biotransport*. Salmon, for instance, pick up PCBs from feeding in the Pacific Ocean, then migrate home to an Alaskan lake, burning fat on the journey. The chemicals build up in their remaining fat and roe, and when they spawn and die, they pass the chemicals to other fish that eat their carcasses. Many animals migrate thousands of miles, carrying the chemicals with them. Wania reports that migratory whales hold voluminous amounts of PCBs and DDT, transporting tons to the Arctic every year.

Perhaps the most unexpected carrier of contaminants to the Arctic has been found on Bjornoya—Bear Island, in the Barents Sea, between Svalbard and the mainland of Norway. The island has been an organochlorine hot spot; its fish, Arctic char, contain about one thousand times more DDT and PCBs than the same fish in northern Canada. Also, compared with nearby Svalbard, remarkably different types of PCBs are found there. What makes the island different? Anita Evenset of the Polar Environmental Centre in Tromso, Norway, thinks she knows. On the island's southern face, where the most contaminated lake is located, a sheer cliff seems to come alive. In actuality, it is a wall of birds. Tens of thousands of seabirds—auks, murres, and kittwakes—sit on its narrow ledges. Glaucous gulls breed atop it. Because the birds can carry a high concentration of chemicals in their bodies, Evenset thinks their feces are the culprit. With so many birds emptying their stomachs there, the pollutants in their waste accumulate in the lake and its fish, upending its food chain and turning it topsy-turvy. Instead of moving from fish to fish-eating birds, the contaminants are moving from birds to fish. Scientists have dubbed it "the guano theory," almost too bizarre to believe.

In the Arctic, mysterious phenomena—whether it's the bird droppings of Bjornoya or the sunrise events that move mercury—have turned out to be the norm.

When it comes to the future of Arctic inhabitants, only two things are certain: Everything we know will change. And no one can predict how.

Where will industrial chemicals and pesticides wind up in the future? Will they increase or decrease? What animals will they accumulate in? All this is up for grabs, dependent on the whims of global climate, which control a complex and unpredictable series of events in the region. Nowhere else in the world are people's food choices and an ecosystem's health tied so strongly to their climate.

Most experts agree that Earth is warming at an unnatural pace. The Intergovernmental Panel on Climate Change, an international body of scientists, stated in a 2001 report that future climate impacts, such as higher temperatures and increased precipitation, are "very likely," while the National Academy of Sciences reports "general agreement that the observed warming is real and particularly strong within the past 20 years." Scientists say that winter temperatures in Alaska, Western Canada, and Eastern Russia have already risen 4 to 7 degrees Fahrenheit in the past 50 years, and they predict that they will increase by another 7 to 13 degrees over the next century, according to the international Arctic Climate Impact Assessment released in the fall of 2004. As global temperatures rise, the higher latitudes are particularly vulnerable to pronounced effects, and substantial changes in the ice have already been documented. Winters are shorter and warmer throughout the Arctic, with temperatures there rising five to ten times faster than elsewhere in the world. Its glaciers are shrinking so quickly that the United Nations warns that the Arctic Ocean may be virtually free of ice in summertime in fifty years. Using satellite images, NASA-funded research has shown that year-round Arctic sea ice is shrinking by 9 percent each decade. Greenland's ice sheet has been thinning by up to six feet per year in coastal areas for the past couple of decades and as it melts, global sea-level rises. If its ice cap, which is two miles thick in places, melted in its entirety, Greenland would release so much meltwater that sea level could rise by twenty-three feet. There are already signs that the warming has changed the creatures that live in the Arctic. Norwegian scientists have found mussels growing eight hundred miles from the North Pole, on the seabed off Svalbard, when they usually favor the warmer waters of southern Europe.

As temperatures rise, five factors critical to the fate of Arctic contaminants—wind direction, the food web, river flows, ice cover, and ocean currents—are likely to shift, altering where they ultimately wind up. "The routes and mechanism by which (contaminants) are delivered to the Arctic are strongly influenced by climate variability and global climate change," says a 2003 AMAP report on climate change and contaminants coauthored by Macdonald. "These pathways are complex, interactive systems involving a number of factors, such as temperature, precipitation, winds, ocean currents, and snow and ice cover." Macdonald, one of the world's leading authorities on how contaminants move through the oceans, says that many of the climactic phenomena in the Arctic will be nonintuitive. Some chemicals may increase while others may decrease. Some animals may be more threatened, some less so. Some regions will experience drops in contaminants, while others will see them rise.

The fate of organochlorines like PCBs and DDT are the most challenging to predict. How they move in the air and water, where they migrate, when they vaporize, and how quickly they degrade are all determined by temperature. One of the most obvious changes if the ocean warms comes in the volatility of chemicals. As temperatures increase, evaporation increases, too. More contaminants would move up through the water column and volatize into the air, which means they are removed from the Arctic food web. That may sound beneficial, but there are lots of other conspirators at play, too. The temperature rise in itself is the least important of all the effects of global warming. A more critical component is the fate of ice. A temperature change of a couple of degrees can alter whether water is water or water is ice. The Arctic's inhabitants are creatures of the sea ice, which is critical to their hunting, so more melted ice could indirectly rearrange the food web. Many animals would have to switch prey. An extra step may be inserted at the bottom of the food web, or perhaps at the top, and every time a step is added, contaminants would build up to higher levels in the top predators. Even if a tiny zooplankton is added to a food chain, contaminants can increase by a factor of five. Walrus, which feed on bottom organisms in drifting ice, may have to switch to other prey such as seal, which would take them from near the bottom of the Arctic food web to the top. That would make walrus—and the people who eat them—substantially more contaminated. Polar bears need landfast sea ice to hunt ringed seals; without it, they may starve. Scientists say bears are already scrawnier in places like Canada's Hudson Bay, and some predict that the Arctic's entire population of 22,000 bears could become extinct as more ice retreats. Also, when polar bears and other animals have less access to food, their fat deposits are depleted and contaminants concentrate in their remaining fat, exposing the animal to even higher doses. On the other hand, some animals with flexibility in what they eat, such as Arctic foxes and even human beings, may switch to more land-based prey, which are less contaminated. As their diet changes, so will their contaminant loads—for better or for worse.

Scientists used to think that climactic changes came slowly in the Arctic. But in 1990 they learned that the climate can shift abruptly. It all started with a boost in the strength of a condition known as the Arctic Oscillation, which, in turn, caused atmospheric pressure to drop over the North Pole. The Arctic has always experienced periods of high pressure and low pressure, but this was an unprecedented change— rapid and widespread. It altered winter wind patterns, which meant that airborne chemicals blew in different ways. For example, pesticides sprayed in Europe and the eastern United States are now more readily transported across the northern Atlantic to the Arctic, and they reach farther north. The new wind patterns also changed ice drift over the pole. Sea ice melted and retreated. The ice pack shrank and open water increased in summer, the prime hunting time for Arctic animals and people. Scientists debate whether the severity of the shift in the Arctic Oscillation was a natural cycle or prompted in part by human-induced global warming. Changes in ice cover aren't unprecedented; they occurred in the 1800s, and probably contributed to accidents among whaling ships that had grown unwary of ice along the coasts.

The oceans are like giant depositories storing contaminants. Ice cover slows gas exchange, so chemicals are trapped near the surface and are slow to disappear. What happens when ice melts depends on the chemical: Each one behaves differently. HCHs are loaded up in the Arctic Ocean from past emissions, just waiting for a chance to escape. With less ice cover, they would evaporate—flux out of the ocean—disappearing faster from the marine food web. PCBs and mercury, on the other hand, are still falling from the air into the ocean, so less ice means more open water, which allows more of the chemicals to infiltrate marine life. PCBs cling to particles in the ocean, so they sink and are eaten by plankton. No one understands how mercury moves up the food web, but it apparently enters the water in a soluble form and then turns into methyl mercury, which is readily taken up by marine life.

Climate change can also alter the course of ocean currents that carry contaminants, as well as change the volumes of freshwater that flow into the ocean. Inflow from rivers and other freshwater systems, particularly from Russia, are a major source of contaminants in the Arctic Ocean. Russian rivers such as the Ob, Taz, Nadym, Pur, and Yenisey are highly contaminated from industries and agriculture. Many of these rivers used to head straight across the North Pole to East Greenland, but during the high-pressure oscillation of the 1990s, they switched to head eastward along the shelf, flowing toward Alaska and the Canadian Beaufort Sea. Under the old flow patterns, eastern Greenland's ecosystem was directly downstream from the Russian rivers but the change in shelf currents could eventually mean more pollutants in Canada and Alaska. Adding more river water also will change how the ocean is stratified, or separated into layers. Freshwater, lighter than seawater, stays on the surface instead of mixing, so if its volume increases, it will alter where contaminants flow. Also, because ice transports contaminants, shrinking floes mean a change in the movement of chemicals. Ice usually drops contaminants right at the edge, in the meltwater, a critical place for wildlife feeding and Inuit hunting.

There are so many variables like these, all interacting with one another, that it is nearly impossible for atmospheric scientists and oceanographers to accurately predict the future of contaminants, even with sophisticated computer models. All models are wrong, scientists say, and that is particularly true in the unpredictable Arctic. Scientists, Rob Macdonald says, must examine all the possible variables, just as a physician does when assessing a patient's health. A physician cannot take just the blood pressure of a patient and declare him or her healthy, just as scientists cannot measure only temperature to determine how chemicals will migrate. "In these systems, putting a finger on a pulse and measuring a particular thing is not going to give you a complete picture," Macdonald says. No one is even certain how PCBs, mercury, and other chemicals are acting and reacting in the environment today, much less how they will behave decades from now. This lack of understanding about the physical and chemical pathways of contaminants is, at least for now, "an insurmountable hurdle" to predicting the fate of the Arctic, Macdonald says.

Global climate change could already be responsible for perplexing and contradictory trends. Some chemicals, such as HCHs, are declining in the Arctic, while others, such as mercury, are increasing. Some locations are cleaner than they used to be, while others are more contaminated. Some species carry more chemicals in their tissues than they used to, while others carry less. If emissions were falling, wouldn't the patterns be predictable in all animals and humans? Not necessarily. Climate complicates the picture. All this variability could be due to changes in climate, not in emissions. Many scientists say human beings are at least partially responsible for altering the Arctic's climate because the burning of fossil fuels emits carbon dioxide that creates a greenhouse effect, warming the planet's surface. The Intergovernmental Panel on Climate Change in 2001 reported "new and stronger evidence that most of the warming observed over the last 50 years is attributable to human activity. . . . Human influences will continue to change atmospheric composition throughout the 21st century." If they disturb global weather patterns, people are also determining their own fate.

Shifts in temperature and ice melt have altered the course of human history in the Arctic many times. Most anthropologists believe that during the last Ice Age thousands of years ago, as oceans turned to ice and sea level dropped, a strip of land across the Bering Strait was exposed. Siberians used it to cross from Asia into Alaska, in search of woolly mammoths to hunt. The Dorset culture, perfectly suited to the cold climate, evolved around 500 B.C. Then temperatures warmed and ice melted. The Dorset, who some scholars believe hunted from the ice pack and rarely had boats, struggled to survive while the Thule people, who hunted bowhead and seal from kayaks in open waters, benefited. The Dorset culture disappeared, replaced by the Thule, who followed the bowhead and quickly spread throughout Alaska, Canada, and Greenland about one thousand years ago. Their arrival transformed the far North, and today's Inuit are their descendants. Today, bowhead and other whales follow the ice edge, and people follow the whales, so climate remains the critical determinant of the destiny of Arctic people. The region's next great cultural shift, scientists say, could be a mere 5 degrees Fahrenheit away.

Chapter 11

Islands of Sudden Change: The Evolution of the Aleutians

Few places on Earth have changed so much, so fast, as the narrow arc of islands where the Pacific Ocean greets the Bering Sea. Alaska's Aleutian Islands are a dynamic place, ever changing. Fog shrouds them one instant and retreats the next. Hurricane-force squalls—as well as giant earthquakes, tsunamis, and volcanic eruptions—descend with no warning. The sea lashes out, frothing at the shore's craggy peaks. Then, just as suddenly, it turns as placid as a lake. Some of the world's most severe weather is born here, where the cold Bering Sea mixes with the warmer Pacific, and what happens in the Aleutians on a given day can influence the climate of all of North America. The Aleutians' environment isn't supposed to be as capricious as its weather. Ecosystems evolve slowly, measured in millennia, not months. But not here, not now. As sudden and savage as an Arctic storm, some mysterious phenomenon—in all probability tied to past human activities—has transformed this spectacular archipelago in just a handful of years. The evolution of the Aleutians illustrates how much the circumpolar North is in flux, and how quickly a fragile food web can fall into disarray.

On a summer day, when the Aleutians' turbulent seas and legendary winds are still, you can hear a killer whale breathe. But look and listen more closely for the sounds of life reverberating off the fog-shrouded cliffs. Where are the sea lions, fat and happy, napping on the rocks and barking at their pups? Where are the furry sea otters crunching on urchins? What became of the ample king crabs and shrimp, and the schools of silvery smelt? And where are the lush undersea kelp forests that provided food and refuge for fish?

Closer to Russia's Siberia than to Anchorage, the Aleutian archipelago is among the world's most isolated places, boasting more shoreline than California and some of the Earth's richest marine habitat. But this vast sub-Arctic ecosystem has suddenly collapsed. And no one knows why. The ecological shifts in the Bering Sea and the North Pacific are unprecedented in scope and pace anywhere in modern history, and some evolutionary biologists even dare to compare it to the extinction of the dinosaurs. Scientists are exploring many factors—climate change, over-fishing, whaling, pollution, disease, predation by killer whales—that have inevitably played some role in the Aleutians' misfortunes.

In a quest to solve this extraordinary environmental whodunit, an eclectic team of men and women spends each August circumnavigating the islands. Virtually alone in this wilderness, members of this small, dedicated group of biologists have been dive-bombed by eagles, bitten by otters, buffeted by 70-mph winds, rattled by earthquakes, and lost in storms. And each year they return for more, drawn back by the Aleutian paradox.

Tim Tinker is swathed in a bulky orange survival suit, hanging from the bow of a twenty-five-foot boat as it hugs the rugged shore of Adak Island. A brutal storm has just ended, leaving skies so crisp that he can see miles away. Volcanic mountains, set against a blue-satin sky and fog as white as cotton balls, are draped with a luxuriant fleece blanket of moss. The green shines so brightly it seems as if it could glow in the dark. Overhead, a bald eagle soars. Dozens of black and white puffins skim across the surface of the sea, their orange webbed feet splashing the 40-degree water.

"There's some bizarre things going on out here," Tinker says.

It seems an incongruous comment for such a perfect August morning. What could Tinker possibly find wrong here? From his perch on the bow, he lifts his binoculars, training them on the rocky reefs. For eleven straight years, Tinker has counted the Aleutians' sea otters, and what he has seen astonishes him as a scientist and troubles him as a nature lover. Tinker drops his binoculars and turns toward the stern of the boat, holding up one finger clad in ragged wool gloves. Manning a notebook and map, assistant Iris Faraklas dutifully makes a notation: One otter. An hour into the survey, they have counted only five otters and two harbor seals. So few of the animals have been spotted that Faraklas has little to write down. She closes her eyes for a moment. The droning of the boat engine is lulling her to sleep. Abruptly, she stands up. It's her first trip to the Aleutians and she doesn't want to miss anything. These are sights that few humans ever see.

"Scenery's kind of drab, isn't it?" Tinker shouts over the roar of the engines.

"Yeah, and all those damn birds are blocking my view," Faraklas yells back.

Jim Estes, part seaman, part scholar, and the team's leader, is at the helm, piloting the boat into the foggy inlets, cautiously navigating around treacherous, submerged rocks. He squints at the shoreline, deepening the crow's-feet that have settled on his face from so many days at sea studying marine mammals. "Back in the old days," he says, meaning a mere eight years earlier, "we would have seen five hundred otters by now." A marine biologist with the U.S. Geological Survey in Santa Cruz, California, Estes, now fifty-nine years old, has traveled to the Aleutians for about thirty-five summers to study the world's largest and healthiest population of sea otters. But it wasn't until the late 1990s that he realized that the otters had virtually disappeared right before his eyes. Throughout the Gulf of Alaska and probably the Bering Sea, too, the balance of prey and predator has been upended, a reincarnation so extreme it's called a "regime shift." Waters once brimming with a traditional mix of seals, otters, and king crab are now dominated by sharks, pollack, and urchins. Virtually no creature remains untouched. Estes himself can't even grasp how much the Aleutians have changed in just ten years. No one has ever seen an ecological decline of this magnitude, in such a short period of time, and over such a large geographic area. "This place," he says, "has gone down the tubes."

If sea otters dream, they are surely dreaming about a place like Adak Island, in the middle of the Aleutian chain. From all appearances, the marine habitat looks ideal. There's plenty of food. Plenty of sanctuary. But by the end of the day, Estes's team tallied the disappointing numbers: They traveled for thirteen hours around Adak on this particular day, surveying more than 200 miles of coastline, but counted only 171 adult otters and 29 pups. One otter per mile. The archipelago is so vast and has so much ideal coastline that it could hold as many as 115,000 otters. In 2003, only an estimated 3,300 otters were scattered across the Aleutians, a phenomenal 96 percent decline since the 1980s, when as many as 100,000 inhabited them. On Attu Island, the island closest to Asia, the population had been robust and thriving for decades, growing by 17 to 18 percent per year. In 1989, 2,000 otters lived there. By 2003, there were only about 100 left.

Estes and Tinker, of University of California at Santa Cruz, first began to notice fewer otters in the Aleutians in the early 1990s but it wasn't until the summer of 1997 that it dawned on them that what they were seeing was not just a temporary blip. In the years that followed, they documented that otters were vanishing from the entire Aleutian chain. Based largely on their work, the region's otters will soon be protected by the federal government. In early 2004, the U.S. Fish and Wildlife Service proposed to list southwest Alaska's otters on the nation's endangered species list. But it is likely too late. A decade ago, eight out of every ten of the world's sea otters lived in the Aleutians. Now so few remain that Estes already considers them virtually extinct. "These animals will be gone in ten years; I mean gone—none left," he says. Like bald eagles, bears, and salmon, otters are a beloved Alaskan icon, the teddy bears of the sea. Even Alaskans who care more about salmon than otters miss the furry creatures. They are the original come-back kid, nearly hunted to extinction by the early twentieth century for their soft, luxurious pelts. When hunting was banned, they re-bounded, thriving by the mid-1960s. This time, though, the otters seem to have encountered a predator even more deadly than men with guns.

At first, Estes, who is the world's leading expert on otter behavior and the role of top-level predators in marine ecosystems, was baffled by the sudden decline. He and Tinker looked for signs of disease, fam-ine, or reproductive troubles, and found none. If thousands of otters had died, where were the bodies? Then it dawned on Tinker: Perhaps the animals are being eaten alive. Killer whales. Estes was incredulous at first. For the voracious, skilled predators, otters are little more than hairballs; snack food for a killer whale, which prefers fatty seals and sea lions. Still, Estes remembered spotting an occasional killer whale, also called orca, lurking close to shore over the years. The sightings, though, were fairly rare. How could just a scattering of whales inflict so much damage? Estes decided to test the theory. They packed up a dead otter on Adak, flew it to California, and ground it up in a blender. Then they calculated its calorie load and compared it to how many calories a killer whale consumes. It turned out that fewer than four whales—3.7, to be exact—could have eaten enough otters around the Aleutians to wipe out 40,000 of them in five years. "We were abso-lutely blown away," Estes says.

No one had ever heard of orcas preying on otters—until the sum-mer of 1992. USGS scientist Brian Hatfield, a member of Estes's team, was on the Pacific side of Amchitka Island, counting otters from the shore. He heard a splash and saw a big male killer whale speeding to-ward an otter two hundred yards offshore. The otter swam into kelp, shaking an injured leg. In the open water, the otter wouldn't have stood a chance, but it hid in the kelp and finally the whale retreated. Since then, the research team has witnessed nine other attacks around the Aleutians. When otters were still plentiful, the occasional attacks seemed like freak events to Estes and Tinker. Now, they say, it appears to be the only reasonable explanation for the disappearance of the Aleutians' otters. Some of their colleagues say there is no real, tangible evidence to prove the theory, although they agree that it is certainly plausible.

It may seem natural, part of the circle of life, for otters to fall prey to orcas. But orcas usually won't touch otters. They had lived in harmony with otters for thousands of years on the Aleutians. Why, all of a sudden, were the orcas preying on them? For killer whales, it was a dietary renaissance. The answer to that question came fairly easily for Estes and other biologists. They simply had to follow the food chain. Orcas customarily feed on whales, sea lions, and seals, which are packed with high-calorie blubber. But the population of Steller sea lions, the world's biggest sea lions, took a precipitous dive in the late 1980s, with only a few hundred left to count during a 1994 survey of the Aleutians. Harbor and fur seals also declined at a similar rate. And the giant baleen and sperm whales were nearly hunted to extinction by commercial whaling operations in the North Pacific between 1946 and 1979. By the time Hatfield saw the orca attack in 1992, otters were the only plentiful marine mammal left in the sea. So they believe the orcas, in their hunt for calories, were forced to switch prey. "How likely is it that all of these things are unrelated to each other?" Estes asks. Instead, he thinks it is part of a "trophic cascade," initiated by the whaling, that tumbled down the region's food web.

It didn't end with marine mammals. With no otters around to eat them, sea urchins exploded in the mid-1990s. Their numbers increased eightfold within a few years. As many as one hundred of the spiny creatures cover each square foot of ocean floor around the Aleutians. The urchins, in turn, have eaten the kelp. In 1993, kelp forests were twenty feet tall and so thick they clogged the engines of Brenda Konar's dive boat. When Konar, a benthic biologist with the School of Fisheries and Ocean Sciences of University of Alaska at Fairbanks, returned a few years later, she couldn't believe the transformation. Only patches of kelp remained, scattered near the shore. In the deepest spots, the kelp was only three feet deep. In a span of fewer than five years, a seafloor layered with a diversity of life had turned into a barren, uniform carpet of green urchins. When the leafy undersea forests vanished, so did many rockfish, snails, starfish, and other creatures that used them to eat and breed. Some populations of seabirds, mainly puffins and kittiwakes, also are hurting from lack of fish. The Aleutians offer proof that one small ecological change can move like a tsunami throughout the entire ocean realm, even if there are no apparent links between species like otters and puffins. Whether bird, fish, or mammal, their fates are intertwined.

Yet the snarl in the food web had to begin somewhere. Scientists wondered: Where? And, even more importantly, who—or what—caused it? Ecological shifts as sudden and sweeping as the ones in the Aleutians can come only from human interference. Sea otters have occupied the Earth for the past 30 million years. Disappearing from their Alaskan stronghold in less than a decade is, at the very least, unnatural. The animals and plants have evolved with no defensive strategies, which gives weight to the theory of human-caused disruption. In pace and in scope, the changes in the Gulf of Alaska and the Bering Sea compare with extinction of the dinosaurs. Yet there was no single catastrophe, no hellish fireballs to blame for this ecological epoch.

Bruce Wright was working for the state of Alaska, managing commercial fisheries, when he witnessed a transformation that puzzled him. All of a sudden there were no crab and shrimp for the fishermen to catch. This was the late 1970s, a time before trawling, so it was unlikely that the commercial fishing industry had wiped them out so quickly. Indeed, species like capelin that had never been fished were gone, too. Something had happened, and it was drastic. Wright knows ecosystems shift naturally—but not like this. "I believe the Gulf of Alaska has seen changes like this in the past, but they are happening faster now, and it's more severe," says Wright, who headed a research division at the National Marine Fisheries Service. "This total regime shift of a large marine ecosystem is new. And it's also well documented. Anybody that comes to Alaska says, 'My God, this place is beautiful.' They say, 'Look at all the puffins and look, there's a bald eagle.'

Superficially, it's a pristine, incredibly breathtaking, beautiful place to be. It looks perfect, like paradise. We would never have known about the changes if people like Estes hadn't been looking. The habitat is still there in the Aleutians. There's not a bunch of condos on the beach but the consequences of human activities are taking their toll, and unless you're conducting long-term monitoring, you won't see those changes day to day. This is a biological regime shift, the consequences of which are astounding. It underscores the fact that everything is connected in ecosystems."

Some theorize that ground zero came in 1977 in the waters of the Gulf of Alaska. That year, the ocean's average temperature warmed by two degrees Celsius, or 3.6 degrees Fahrenheit, and has since stayed that way. The Arctic has been especially vulnerable to climate change. At the North Pole recently, Russian researchers saw ten square miles of open water where they had always seen ice. Going back in time to prove their theory is impossible. But this, marine ecologists say, is a possible scenario of the chain of events that happened beneath the surface of the sea: Plankton was probably the first thing to disappear because it is ultrasensitive to temperature changes and lives less than a year. Tiny copepods and krill probably followed quickly. Then, finding their food gone, the shrimp and crab, along with smelt fishes such as capelin and herring, vanished, replaced with an explosion of cod and pollack and sharks. Once-thriving shrimp and crab fisheries collapsed while large fishing trawlers descended on Alaska, harvesting millions of tons of pollack and cod a year. Whether this contributed to the disappearance of the marine mammals is disputed. Some biologists subscribe to a "junk food hypothesis," suspecting that the absence of the smelt fishes, which are high in fat, might have contributed to the decline in seals and sea lions because the baby mammals could not find enough calories to survive the winters.

Estes's theory, however, is entirely different. He discounts the climate regime shift theory and believes instead that killer whales were left with little to eat after commercial whaling in the 1950s and 1960s, so they preyed on seals and sea lions, decimating their populations.

Other scientists believe that the sea lions are competing with the commercial fishing ships for pollack, cod, and mackerel, so the federal government imposed restrictions on the industry in 2000. A panel of the National Research Council couldn't untangle it all: The experts concluded in 2003 that a variety of forces probably led to the sea lion's decline. For whatever reason, it is obvious that all of a sudden, the Aleutians have turned into a predator pit, unsafe for otters, seals, and sea lions. Everything, it seems, is working against survival of the islands' marine mammals, although so far, the rest of Alaska has escaped the changes.

Scientists have not figured out what role, if any, pollution may be playing in the ecological shift. Pesticides including DDT, mirex, chlordane, and HCHs have wound up in the Aleutians from thousands of miles away. Bald eagles that never leave the islands have excessive concentrations of DDE, a metabolite of DDT, even though the pesticide was never sprayed there. Those nesting on the westernmost island, Kiska, contain among the highest DDT levels found in birds anywhere in the circumpolar north. The contamination increases from east to west, suggesting that the pesticide originates in Asia and moves to the Aleutians via ocean currents and winds. But scientists Robert Anthony and Keith Miles, with the USGS in Corvallis, Oregon, and Davis, California, suspect biological transport plays an even bigger role. Migratory seabirds like fulmars, glaucous gulls, and puffins are contaminated with DDT and DDE. They probably pick up the pesticide in distant waters, perhaps in Asia, then carry it back to the Aleutians, where they are eaten by eagles. Anthony and Miles, in a 2004 report, hypothesized that the disappearance of the otters, in an indirect way, may be responsible for high concentrations of chemicals in the western Aleutians' bald eagles because the eagles are eating more seabirds and less fish due to the collapse of the kelp forests triggered by the otter decline. Eagles on the islands that rely most heavily on seabirds for their diet have the highest DDE concentrations in their eggs, and they have less success reproducing than eagles with lower levels. DDT, which thins eggshells and kills developing chicks, was responsible for nearly wiping out eagles and other birds in much of the developed world in the 1960s. Although no eggshell thinning has been detected in the Aleutian nests, the most highly exposed eagles—those on Kiska Island—do produce fewer chicks. Worrisome amounts of mercury and chromium have been found in some eagle carcasses, too.

PCBs are also found in the Aleutians, in its mussels, sea lions, and otters, although the source is probably local, not global as it is with the DDT and mercury. Otters here contain nearly twice as much PCBs as otters in California and thirty-eight times more than otters in southeast Alaska, according to AMAP data. The source, however, is likely to be the U.S. military, which shut down a base on Adak in the 1990s. Monitoring of mussels in Adak's Sweeper Cove has revealed some unexpected patterns, particularly year-to-year variation in PCBs levels, which suggests, Estes says, that the "movement of contaminants through that system, although localized, is highly dynamic." Why do the levels ebb and flow? And what does this mean for the health of creatures here? Although PCBs can alter reproduction and suppress immune systems, Estes doesn't believe they are responsible for the disappearing otters because he has found no declining birth rates or disease outbreaks. Their effects, if any, remain largely unknown.

On an August day, Walter Jarman stands next to a pile of unrecognizable military debris in a grassy field on Adak. He bends down and picks at the pile of metal and wood, trying to figure out what it is. The islands are the weirdest place he has ever studied. He and Estes are puzzled by virtually everything they find here. Everything seems to defy logic. They concoct logical theories to explain ecological trends or events, only to be proven wrong. The Aleutians are so unpredictable that scientists may never unravel the biological interactions and prove or disprove their theories about how the food web got so tangled. "An unfortunate laboratory," Jarman calls this sub-Arctic terrain. How could a place this spectacular be so screwed up? By sea, Adak appears pristine. On land, it is an eerie place, holding secrets of its military past. Today only a ghost town remains. A McDonald's, the only restaurant on the island, is boarded up. Military houses are empty, their lights off, their grass overgrown, their driveways empty, their swing sets idle. Not a single dog barks. Jarman walks down the middle of the street. No need to look for cars. Unexploded ordnance is a bigger threat.

"It's such a land of contrasts here," he says. "There's something so contradictory about a place this beautiful and remote and contaminated."

In much of Alaska, August is prime tourist season. Sports fishermen from around the world flock to its coastline and lug home ice chests packed with salmon and halibut. But not in the Aleutians. Hardly any tourists venture here. At the gateway of the Arctic Circle, halfway between Tokyo and Seattle, this is the spot where East meets West. It's a four-hour flight from Anchorage to Adak, and from there, it's thirty-eight hours by ship to Attu, which sits 1,300 miles off the mainland on the far-western edge of the chain, next to Russia. It's not easy to chronicle nature in a place so inaccessible, so forbidding. For a month each summer, the research team works out at sea, collecting data from daybreak to nightfall, which in this part of the world stretches for fifteen hours. Along a shoreline known to be treacherous to navigators, Estes and his team log a thousand miles in their twenty-five-foot boat, combing the shoreline for otters.

Estes has weathered just about anything this extreme environment can offer over the past thirty-five years. One night, just before midnight on Adak, Estes and his colleagues are settling into the bunkhouse after spending another fifteen-hour day at sea, harvesting fish, diving for urchins, measuring kelp. As they devour a late dinner of fresh halibut tacos, Estes tells them about his worst day in the Aleutians. It was an August afternoon on Attu Island. Their boat engine had failed and an unexpected, winterlike storm had hit. Lost in the fog and freezing temperatures, Estes and two other scientists hiked for nearly three days, covering maybe a hundred miles. One, a student researcher and a triathlete, suffered hypothermia and Estes was forced to consider abandoning him to die. They all found their way back to camp, alive due to luck, not skill. Estes recalls the experience calmly, with no trace of hyperbole, recording his personal ordeal as matter-of-factly as he does his scientific data. Still, the other, younger researchers are transfixed by his tale. It is a harsh reminder that their wild laboratory is an unforgiving place, even in its mildest month. They tidy up the kitchen and retreat to their bunks, just as the midnight sun is setting.

The plight of the Aleutians—and the High Arctic—reveals that proximity to people is not always the best indicator of environmental damage. There are no clear-cut forests. No rows of red-tiled roofs. No industrial smokestacks. This wilderness still looks as nature intended. And there lies the paradox. It looks wild. It feels wild. But with such unnatural events, can it still be considered wild? The Aleutians are one of the most sparsely populated places on the continent. On Adak, an island eleven times bigger than Manhattan, the last census counted 316 people. The entire 1,100-mile chain is inhabited by only 8,000 humans—and more seabirds than the rest of the United States combined. No wonder what happens on this archipelago goes largely undetected. In 1986, the largest earthquake recorded in North America struck the Aleutians. But hardly anyone was around to notice. These islands also were one of only two U.S. locations to be attacked during war. Everyone remembers Pearl Harbor. But who remembers Attu and Kiska? The islands were invaded by the Japanese and fortified by the U.S. military, but few people know anything about them. In a letter home, an unmarried soldier stationed on Adak Island during World War II reassured his worried mother that "there is a girl behind every tree." She, most likely, was unaware that the Aleutians have no trees.

Ironically Estes initially was drawn to the Aleutians by its bounty of marine mammals. He never expected to witness its ecological transformation. Now he's haunted by a suspicion that other ocean realms around the world are undergoing similar drastic changes—and no one is around to see it happen. If this robust, remote place is collapsing, where on Earth is safe?

One thing is certain. The Aleutians, like the rest of the polar region, will keep evolving. Some experts say the ocean there seems to be cooling and shifting back to the temperature it was before the sea lions began to vanish. But even if all the elements—the otters, the seals, the sea lions, the kelp—were to return, these islands will never be the same. Like marble chips in a kaleidoscope, they would fall in different patterns.

Estes knows he won't see the Aleutians recover in his lifetime. He is despondent that all the otters there likely will be gone—extinct—in a few years and he knows he is helpless to stop it. Between 2000 and 2003, even after he warned of the islands' ecological collapse, the otter population plummeted by 63 percent. He has grown frustrated by scientists' inability to get things done, troubled by their failure to find solutions to dire ecological problems and influence environmental policy. He thinks his colleagues should work toward consensus, since certainty will always elude them. What, though, can be done to help the Aleutians' otters? Restrict fishing? It probably won't do a thing. Cull killer whales? Not too likely. The ultimate solution, he says, is to "let the whole ecosystem recover," to let time heal the wounds inflicted upon it.

From the three decades of snapshots he carries in his head and the three decades of otter data stored in his computer, Estes knows wilderness is an illusion here. Yet when he steers the boat past spidery waterfalls spreading like webs down a sheer cliff few human eyes will ever see, the Aleutians still feel wild to him. And that sensation, decidedly unscientific, gives him a glimmer of hope. He will be back next year.

PART III

SOLUTIONS AND PREDICTIONS

Chapter 12

The Diagnosis:
Scientists Write a Prescription

It's mid-January 2002, far above the Arctic Circle on the edge of the Norwegian Sea, and about two hundred scientists from around the world are gathered at the Polar Environmental Centre in Tromso, Norway. Their mission over the next three days is to decide whether contaminants are jeopardizing the health of the Arctic's animals and people, and present their findings in the fall to the environmental ministers of the eight Arctic nations—Canada, Denmark/Greenland, Finland, Iceland, Norway, Russia, Sweden, and the United States. To go behind the scenes at this meeting is to get a glimpse of how scientists struggle to send a cohesive message amid uncertainty. This is a select group, some of the premier experts from throughout Europe, North America, and Asia, propelled by their scientific curiosity and their love for the far North. Since its creation in 1991, this international collaboration—the Arctic Monitoring and Assessment Programme—has been critical to discoveries related to contaminants in the Arctic environment. It has meant an intense, focused infusion of science, turning the Arctic into a laboratory for biologists and chemists studying the effects, sources, and pathways of toxic chemicals. Their work reverberates far beyond the northern latitudes, since the effects of contaminants have global implications. AMAP's first report was presented to the ministers in 1997, and since then, scores of researchers have updated, refined, filled in the voids, trying to make sense of it all.

Johan Mikkel Sara, a member of the Saami Parliament of Norway, representing the nation's 50,000 indigenous people, appears at the Tromso meeting dressed in a traditional Saami red and blue embroidered coat. He welcomes the scientists, reminding them why they are there by expressing the frustration and fears of all Arctic peoples and appealing to them to find solutions. "With regard to the challenges associated with global conditions, there is not much indigenous peoples in isolation can do," he says. "I am referring to the long-range air- and waterborne transport of pollution from other parts of the world. Yet the problems affect the very foundation on which the lives of indigenous people are based. Indigenous people's way of life and their strong cultural ties to nature make them particularly vulnerable to pollution and other conditions that impact the balance between people, animals, plants, and the land."

Sara ends his talk with a poem from a Saami writer.

My home is in my heart. It migrates with me. What do I say to strangers who spread out everywhere? How shall I answer their questions that come from a different world? . . . I say nothing. I only show them the tundra.

Outside the window, the northern lights are glittering in the dark sky over Tromso, arguably the Arctic's most beautiful and cosmopolitan town. The morning sky begins to brighten. By 11 A.M., the sun has risen over the mountains for the first time in two months. A few hours later, after lunch, bands of dark blue stretch across the sky like ribbons, signaling the end of the first day of polar sunrise and the beginning of another long night—twenty hours of it. By 4 P.M., it's pitch black again in Tromso but the scientists inside the polar institute are too preoccupied to notice.

Derek Muir, a tall Canadian with a trim beard, wearing a tweed wool jacket and wire-rimmed glasses, steps to the podium after Sara. He's a detail man, capable of noticing all the fine points while still seeing the entire picture. For years Muir, of Environment Canada's National Water Research Institute in Burlington, Ontario, has been known as a human clearinghouse for Arctic data. Toxaphene in ringed seals in the White Sea? PCB congeners in polar bears in Svalbard? Flame retardants in Nunavut's belugas? He knows it all, yet he's keenly aware of how much he doesn't know. To substantiate theories about what is happening in the Arctic, he needs to document trends spanning decades and thousands of square miles. A single high reading is dubious; a pattern of them offers more certainty. He doesn't want a snapshot in time or space—he wants a running, decades-long documentary for the entire circumpolar north. He worries particularly about the lack of wildlife data from Siberia and the Russian far east—a big, gaping black hole for validated research, although Russian officials, funded through AMAP and assisted by Canadians, have begun to release data from the monitoring of people and animals there. Muir's dream is finding comparable data from throughout the Arctic, year after year, for all species of animals and plants.

But for now Muir has half an hour to sum up everything he knows. "It's a huge subject," Muir says, apologizing to the scientists for the shorthand of his morning session. Muir starts with what he calls the "dirty sixteen" or the "legacy organochlorines," the contaminants such as DDT and PCBs, most of them banned decades ago. He knows that the Arctic has become an important global indicator—a canary in the coal mine—for how these chemicals persist and accumulate in nature. Several hundred Arctic scientists have been trying to answer important questions: With the bans in effect, are the contamination levels dropping? Which animals and places are still at risk? Are there new chemicals of concern? Where is mercury increasing, and why? Many of these scientists work with paltry funds in grueling, subzero conditions, where sometimes just reaching the animals you want to study means taking your life in your hands.

Muir, accumulating data from a variety of international scientists, has some answers but lacks others, and mercury, in particular, remains a mystery. He tells the group that there is "great variation" in the mercury levels of animals throughout the Arctic, leaving a scrambled puzzle with too many missing pieces to solve. In western Greenland, mercury levels in seals are low but increasing while in Canada they are high and stable. Along the Beaufort coast, mercury in beluga is rising dramatically while in the beluga of eastern Canada it is staying the same. Concentrations of the powerful, neurotoxic metal are greatest in polar bears, glaucous gulls, and Greenland sharks, and the levels found in seabirds such as fulmars, murres and kittiwakes are growing, indicating that the Arctic is continuing to serve as a giant "sink" for globally migrating mercury. Over the long term—the past few decades—most species contain more mercury than they used to, but over the short term half the data show no change while half show an increase. It's an indecipherable picture for Muir. Why is mercury worsening in some places and in some animals but not in others? The trends are almost as perplexing for the pesticides and PCBs. While nearly all areas have been declining at a slow pace over the past few decades, hot spots remain. Muir voices his suspicion that Russia is a continuing source of new PCBs found in the ringed seals and polar bears around Svalbard. Then he moves to what he calls the new legacies— chemicals that are starting to appear in Arctic animals. He ends with some predictions: The old PCBs and DDT will decline in much of the Arctic, but other chemicals will magnify.

A parade of scientists follows Muir, discussing for hours their specific findings of tests on a virtual Arctic zoo—sculpin, cod, char, beluga, killer whales, kittiwake, fulmar, ringed seals, fur seals, zooplankton, foxes, bears. They have measured dioxins, furans, PCBs, chlordane, DDT, HCHs—dozens of chemicals—and checked the animals for an array of effects—changes in sex hormones, thyroid hormones, retinol, immune cells, birthrates. In just a four-year span, since their first report was published in 1997, AMAP has generated reams of new data.

After a day focusing on wildlife, the scientists turn their attention to another top predator—human beings. They talk in great detail about what they have found in people's cells, hair, breast milk, umbilical cord blood, and what it seems to mean for their immunity, estrogen and testosterone, infectious disease rates, cardiovascular changes, and even bone density. They investigated what and how much food Arctic people eat and what contaminants it carried, and analyzed it for any link to these effects. They agonized over the details. Do the omega-3 fatty acids and antioxidants in the food protect people against the harm of the contaminants? Are the chemicals causing ear infections in Nunavik children? Is the mercury in the Faroes contributing not only to brain changes but also to heart disease and osteoporosis? And what surprises can we expect from this suite of new chemicals emerging in the Arctic?

Only a handful of scientists from the United States are attending this international brainstorming session. In fact, there are more from Russia than from the United States. For a decade, Canadian and Scandinavian scientists have dominated most of AMAP's work, while the United States has taken a backseat role in exploring contaminants in the Arctic. The reasons are manifold, but a major one is the lower levels of contamination found in the western Arctic. Most chemical pollutants flow to Canada and Greenland, rather than to Alaska, and Alaska's Inupiat eat bowhead, which is much less contaminated than the narwhal, seal, and beluga consumed in Greenland and Canada. Still, there are important contaminant problems in Alaska, including chlordane in bowhead, DDT and PCBs in the Aleutians, and mercury in fish caught by the Yup'ik in the Yukon delta. Scientists say there is a historic explanation for the United States' lack of leadership over the past three decades. In the 1980s, American researchers focused on another Arctic issue: They thought radioactivity from Soviet nuclear weapons tested during the Cold War would be the issue of overriding importance to the health of indigenous people. Meanwhile, Canadian scientists took another path, concerned about PCBs and organochlorine pesticides instead. High radiation levels were found in some areas, but they have declined and they are minimal in most of the Arctic. The Canadians' hunch panned out, and American researchers never caught up, although their involvement in AMAP has grown in the past few years.

After three days of information overload, the scientists divide into groups and retreat into workrooms to begin distilling it all for the political leaders of the eight Arctic nations. Cynthia de Wit, a no-nonsense, straight-shooting scientist from the Institute of Applied Environmental Research at Sweden's Stockholm University, is leading the group directed to reach conclusions about chemicals in wildlife. Seventeen biologists have gathered in the room. As scientists always do, they begin with a question. Have we documented biological effects in wild animal populations? "Who wants to start?" de Wit asks. Polar bears, the icons of the Arctic, kings of all predators, have been the focus of the most concern and debate.

They have just spent days poring over each other's data and analyses and now they must reach a consensus, seeking order out of all this chaos. What does it all mean? Are Arctic animals in danger? What specific evidence do we have of that? Next door, their colleagues are wrestling with the same questions about human exposures in the Arctic. Years worth of data collection and tons of detailed findings will be condensed into a few sentences that the ministers will read and the public can understand. Environmental scientists tend to dwell on what they *don't* know, but this time their job is to agree on what they *do* know. This is agonizing for experts skilled in recognizing subtleties and trained as skeptics, people who typically answer each question with another question. By their very nature, they are never satisfied that they know enough to make a bold or sweeping statement. Ask this crowd if the sky is blue and someone is bound to ask: What do you mean by sky? What time of day? What season? What altitude? What shade of blue? Then they would set to work developing a battery of tests to answer the simple question. What looks like confusion and uncertainty among scientists is really just a preoccupation with being cautious. These are the people that the rest of the world depends on to find the truth, to guide governments' policies and laws. The questions that AMAP wrestles with in the Arctic are a microcosm of all the world's environmental challenges. Before making the hard decisions that can alter people's health and livelihood, the public wants to hear some semblance of what experts consider the truth.

An hour into their discussion, the scientists have debated many facets but they have still failed to come up with a simple answer to their first question: Are there significant biological effects from contaminants in the Arctic? The answer is yes, of course, but they refuse to say so until they debate all the nuances and uncertainties. "When it comes to biological effects, we need to make sure we can stand by whatever we say," de Wit warns the group. It's difficult, sometimes impossible, to prove that a contaminant has harmed a population of animals. Sometimes scientists can't even tell the effect on an individual animal, much less an entire population. There are few examples of success in this, even now, and that remains the biggest challenge for AMAP.

One scientist mentions that there are no data to show that populations of polar bears, gulls, foxes, or any highly exposed Arctic animal are declining due to contaminants. The others agree, but also note that there is likely to be a long time lag between a species' exposure and its decline—perhaps twenty years. Once a population starts to disappear, it might be too late to save it. De Wit comments that the immune changes they have found in animals pose a real threat because if a disease moves into an area, the population would plummet. While they definitely found correlations between effects in animals and chemicals in their bodies, they cannot prove causality. Causality is the bogeyman of environmental science, a continual weak point. Research on animals in the wild is not a controlled experiment. Scientists cannot dictate the circumstances of how wild animals—or humans—live as they can with lab mice, so they can never indisputably prove that contaminants, rather than some other unknown factor, cause a specific effect. The best they can do is document the same correlation so often that it is unlikely to be a coincidence. Making matters worse, no body contains just one contaminant. Each carries a brew of them, hundreds of different compounds, and these scientists know that they aren't even close to understanding the synergistic effects.

The group breaks for the night and returns the next morning, refreshed and more focused on its mission. One scientist warns: Let us not be too timid, too weak. We found effects, let's say so. Our message is strong, de Wit agrees, "a lot stronger than we think." Their ultimate answer to the question is yes: We have documented significant biological effects from contaminants in the Arctic.

Nine months later, after many more phone calls and rewrites, AMAP's final report summarizing the scientific findings is made public in October 2002, at a meeting of the ministers in Finland. It is the first of thousands of pages of details to come in the AMAP report. "Adverse effects have been observed in some of the most highly exposed or sensitive species in some areas of the Arctic," the report says in a summary of the impacts on wildlife. It goes on to list many animals—polar bears, peregrine falcons, glaucous gulls, northern fur seals, char, and dogwhelks, among others. It documents a list of specific effects, hedging some as far as their meaning, and urging more research. Scientists fear making sweeping statements. They worry that mistakes might be made, that money might be misspent. But they are relieved that at least the paralysis over the Arctic dilemma will be broken. Their message, as de Wit predicted, is clear: The Arctic is in danger. The diagnosis isn't good. Now it is the politicians' job to prescribe solutions.

Chapter 13

POPs and Politics:
Taking the First Step
Toward a Solution

Chemicals cross borders en route to the Arctic, so the solutions must, too. Nevertheless, international consensus when it comes to environmental matters does not come quickly or easily. It is a long, painful process. During three years of negotiations over how to phase out twelve of the planet's most dangerous contaminants, nations squabbled, mired in legal details. Rising above the fray, the most inspiring voice came from a proud and eloquent Inuit woman born in the tiny Canadian village of Kuujjuaq.

In the early 1990s, as scientific evidence emerged that DDT, PCBs, and other decades-old compounds were still poisoning people and animals worldwide, pressure was mounting for international action. The United Nations Environment Programme (UNEP) stepped to the forefront in 1995, adopting a directive called Decision 18/32, authorizing a review of global policies related to the so-called Dirty Dozen. All twelve chemicals—nine of them pesticides—are persistent organic pollutants, or POPs, that are toxic to people and wildlife, accumulating in the environment and transported globally via the air and oceans. Two years later, the UN group announced that the chemicals posed serious risks and recommended an international treaty to govern their use.

Negotiations commenced in July 1998. Delegates from more than one hundred nations gathered in Montreal. For many of them, living in industrialized nations with a multitude of pollution problems, the POPs issue seemed remote, impersonal, even a bit arcane—until a proud and articulate Canadian Inuk named Sheila Watt-Cloutier, representing the 155,000 Inuit inhabiting four Arctic nations, stepped to the podium. She told the delegates about her heritage, a youth spent at Ungava Bay in Nunavik eating caribou, fish, whale, and other traditional foods gathered by her brothers. As president of the Inuit Circumpolar Conference, she pleaded with the UN delegates to protect people in the Arctic—and everywhere—by ridding the world of the poisons contaminating their foods, their environment, their bodies. Her people, she said, are the globe's early warning system for POPs. "A poisoned Inuk child, a poisoned Arctic, and a poisoned planet are all one and the same," she said. The audience at one forum during the session gave her a rare standing ovation. Many participants say her impassioned and poignant plea gave their mission a human face and instilled in them the will to persevere with negotiations. Never before had the Inuit commanded so much attention in a global arena.

Six months later in Nairobi, at the second round of negotiations, UNEP's executive director, Klaus Toepfer of Germany, announced that POPs must be stopped and a global solution must be reached by the following year. Watt-Cloutier, impressed with the strength of Toepfer's convictions even though his homeland had no Arctic people of its own, presented him with a soapstone carving of a Canadian Inuit mother cradling a baby. He in turn presented it to John Buccini, chair of the negotiations, who vowed to display it at the head of the table for the remainder of the talks. For the next year, it served as a powerful symbol of the need to protect future generations from PCBs, DDT, and other contaminants that mothers, particularly in the Arctic, pass to their children. When negotiations were long and tense, and some delegates started arguing about economic ramifications and the need for insecticides to control pests, Buccini said he only had to touch the Inuit sculpture for strength and clarity. Safer alternatives, he concluded, must be found.

That fall, at the third session in Geneva, negotiations started to split between the north and the south. Tropical nations, particularly in Africa, wanted to keep using DDT to control malaria-infested mosquitoes or risk thousands of people dying of the disease every year. Battle lines were being drawn—but they vanished when Watt-Cloutier announced that the Inuit would not accept any treaty that pitted their health against the health of other indigenous peoples. "I cannot believe that a mother in the Arctic should have to worry about contaminants in the life-giving milk she feeds her infant. Nor can I believe that a mother in the south has to use these very chemicals to protect her babies from disease. Surely we must commit ourselves to finding and using alternatives," she said in her speech.

Finally, almost two years later, on May 22, 2001, the pact—named the Stockholm Convention—was completed. The ministers of ninety-one countries and the fifteen-member European Commission signed it, agreeing to ban or severely restrict the chemicals and establish a process for adding new compounds to the list. Limited use of DDT would be allowed in some tropical nations for malaria control on a strictly regulated basis. The signatory nations pledged $500 million in support for eliminating the Dirty Dozen, including efforts to develop safer pesticides and to locate and dispose of old stockpiles in developing nations. At the top of the thirty-four-page pact, the UN delegates acknowledged that Arctic ecosystems and indigenous people were particularly at risk from POPs—they were the only communities singled out.

"The Stockholm Convention will save lives and protect the natural environment—particularly the poorest communities and countries—by banning the production and use of some of the most toxic chemicals known to humankind," Toepfer announced.

The next step was ratification, which allows each country to implement the treaty. Canada, Watt-Cloutier's homeland, became the first nation to ratify it, acting immediately after it was signed. In the 2003 book *Northern Lights Against POPs,* she recalls her pride in her nation— a nation in which her people had always felt like foreigners—and her restored faith in the compassion of the non-Inuit world. She told the environmental ministers that the treaty "brought us an important step closer to fulfilling the basic human right of every person to live in a world free of toxic contamination. For Inuit and indigenous peoples, this means not only a healthy and secure environment but also the survival of a people. For that I am grateful. *Nakurmiik*. Thank you."

The United States, long an absent player in Arctic contaminant issues, embraced the pact at the convention in Stockholm. Christine Todd Whitman, then administrator of the U.S. Environmental Protection Agency, signed it, announcing that President George W. Bush was personally supportive of it. Whitman said the United States was already on its way to implementing its terms—it had already banned or restricted the twelve compounds and was cleaning up old contaminated sites—and she guaranteed financial support for developing nations to find safer alternatives and deal with their stockpiles. Soon afterward, Bush pledged his support in a Rose Garden ceremony. "We must work to eliminate or at least to severely restrict the release of these toxins without delay," the president announced.

Three years later, in May 2004, the POPs treaty became law in the fifty-nine countries that had ratified it by then, including Canada, Mexico, Japan, many African and Asian nations, and all major countries in Europe. Yet the United States, despite Bush's earlier assurances, was noticeably absent. Despite widespread agreement in the United States, even from pesticide and chemical companies, for the provisions related to the original twelve chemicals, Congress and the president were unable to agree on terms for adding other compounds in the future. For the United States to become a full partner in the treaty, Congress must pass legislation that amends two long-existing federal laws, the Federal Insecticide, Fungicide and Rodenticide Act (FIFRA) and the Toxic Substances Control Act (TSCA). Those bills remain mired in political and legal disputes over whether the U.S. EPA can automatically enforce restrictions on new compounds after the nations collectively agree to treaty additions or whether it must launch a more rigorous economic review before enforcing them. As a result, the pact was unenforceable in the United States when it became law on May

17, 2004. Of the eight Arctic nations, only the United States and Russia have failed to ratify it.

Still, the day that the Stockholm Convention became enforceable in the fifty-nine nations was considered a day of celebration among Inuit leaders. They decided there was no better way to celebrate than to gather with Buccini for a feast of seal, caribou, whale, and char— the very foods contaminated with the chemicals facing the new restrictions. Still, Watt-Cloutier warned her people not to get complacent. As hard-fought as the Stockholm Convention was and as vital as it is to controlling global contaminants, its passage is merely a first step in cleansing the Arctic.

New toxic threats continue, unabated. An estimated 100,000 chemicals are in commerce today, and toxicologists say that little or nothing is known about the hazards of most of them. One in every five high-volume chemicals lacks even basic toxicity data, while only 14 percent have good data, says Finn Bro-Rasmussen, professor emeritus of Technical University of Denmark. He estimates that almost half should be classified as hazardous. "A lot of chemicals are on the market, and people think they have been tested but often they haven't," says Stockholm University's Cynthia de Wit. "It's rather scary that the data sets are so poor."

Europe is taking additional steps to clean up its own act. In October 2003, the European Union's executive branch, the European Commission, proposed a bold new policy called Registration, Evaluation, and Authorization of Chemicals (REACH), which would fundamentally alter the way that chemical compounds are regulated by government and tested by industry. Under the draft law, companies would have to register basic scientific data for about 30,000 chemicals with a newly created European agency. Of those, chemicals used in the largest volumes and those already linked to health or environmental hazards would be subjected to additional testing and possible bans.

If adopted by Europe's Parliament and the Council of Ministers, REACH will be the world's most comprehensive regulation governing the use of chemicals. The European Commission, representing fifteen nations, crafted the proposed policy because of growing concern over an array of chemicals contaminating humans and wildlife.

Under current laws, only chemicals that were first used after 1981 in Europe and after 1976 in the United States must undergo testing for environmental risks. The new law would regulate all of the estimated 30,000 chemicals used in volumes exceeding one ton per year in Europe and require basic safety testing of those used in excess of ten tons. The most stringent rules would be aimed at about 4,500 compounds used in larger volumes, over 100 tons per year, and at least 1,500 compounds of "very high concern" to the EU because they are known to cause cancer or birth defects, to build up in bodies, or to persist in the environment. Manufacturers of chemicals with known health or ecological risks would need government authorization, extensive testing, and proof that the benefits outweigh the risks—similar to the review necessary for pharmaceuticals. Many of these compounds are widely used—from benzene found in crude oil to flame retardants in computers. The authorization process is the most worrisome part of the proposal for U.S. industries, since EU officials estimate that 300 to 600 compounds would be withdrawn from commerce. Parliament could mount its first vote on the policy as early as 2005.

Europe's proposal faces powerful opposition from industry, particularly in the United States, which exports about $20 billion in chemicals to Europe annually. Industries in Europe and the United States say the regulations would be costly and unwieldy. Some of the strongest opposition comes from the Bush administration, which has sided with the American Chemistry Council, a group representing industry, and tried to rally support from EU member nations and its trade partners in Asia and North America to stop REACH because of the cost to chemical industries. Bush's then Secretary of State Colin Powell lobbied against it, saying that a $9 billion U.S. industry was at stake. The European Commission estimates the direct cost to the chemical industry at $2 billion over an eleven-year period, while industry groups predict that it would be many times higher. Experts say that in the United States such a chemical policy has little chance of winning congressional support because of the impact on industries with powerful lobbies. Nothing like REACH has ever been proposed by any U.S. president, whether Democrat or Republican. It is harder than ever, experts say, to get a chemical banned in the United States.

That disturbs many environmental scientists. Canadian atmospheric scientist Terry Bidleman has written that a chemical's presence in the Arctic should set off alarm bells because it offers conclusive evidence that it is persistent in the environment and transported globally, posing a worldwide threat. Toxicologists say they are virtually certain that other chemicals in use today are just as troublesome as PCBs and DDT. Some suggest that government regulators restrict all compounds that meet certain chemical properties, such as long half-lives or high volatility, that make them prone to build up in the environment and wind up in the Arctic. Bans on the Dirty Dozen aren't enough, they say. All chemicals that tend to evaporate and survive long enough to make a long journey should be regulated, too, Donald Mackay says. "Further action may be necessary to assure that POPs are not just replaced with substances with the same undesirable global distribution behavior," Mackay and Frank Wania wrote.

Mercury, one of the most dangerous contaminants in the Arctic, is not regulated under the Stockholm Convention or any international agreement. But UNEP is now making efforts to address that. In 2003, UNEP's governing council declared that individual nations should take immediate action to protect human health and the environment from mercury releases. "We live now in the twenty-first century and there can no longer be any excuse for exposing people and the natural environment to dangerous levels of toxic chemicals," Klaus Toepfer said in 2003, when UNEP's mercury assessment called for reductions in emissions. "In the case of mercury—which has destroyed the lives of thousands of people—we need to make mercury poisoning a thing of the past." Michael Bender of the U.S. environmental group Mercury

Policy Project, and a representative of a global coalition of twenty-eight organizations, says that "no single country can resolve the mercury problem on its own. There are alternatives for most all mercury uses, but there is no alternative to global cooperation."

That cooperation is not likely to happen anytime soon. European delegates are trying to lay the foundation for a binding international pact for mercury similar to the Stockholm Convention, but negotiations will be time-consuming and difficult, generating perhaps even more international conflicts than arose during the years-long dialogue over the Dirty Dozen. Norway and Switzerland, at a special UNEP meeting in February, 2005, advocated phasing out use of mercury in products and manufacturing processes worldwide. For example, nine chlorine-manufacturing plants in the United States and fifty-three in Europe still use an outdated technology that relies upon large vats of mercury to trigger a chemical reaction. So far, however, there is little international support for a treaty that limits use of the heavy metal. The Bush Administration instead is promoting voluntary partnerships between industries and nations, with no specific targets.

In the meantime, mercury levels continue to rise in much of the Arctic. And mercury isn't the only substance that is increasing. There are hundreds, perhaps thousands, of pesticides and other chemicals used in large volumes today that build up in the environment and are readily transported around the globe by winds and waves. One, in particular, has already turned up near the North Pole, and its legacy, scientists fear, could haunt the Arctic for generations.

Chapter 14

The Chain of Evil
Continues Unbroken:
The Arctic's New Toxic Legacies

In southern Sweden, along a river that empties into the Baltic Sea, fishermen in 1979 caught some pike and turned it over to food inspectors for routine chemical tests. The fish, as expected, contained traces of pesticides and other common contaminants. But along with them, the chemists were surprised to see new ones they had never seen before: polybrominated compounds used as flame retardants.

For Swedish scientists, it was the worst form of déjà vu: Tests in Sweden had offered the first clue, back in 1964, that PCBs were pervasive in the world's environment, in wildlife and in human bodies. Could this be a sign of a new threat? Published in 1981, the report on the discovery of flame retardants in the river caught the eye of Åke Bergman, chair of environmental chemistry at Stockholm University and one of the world's foremost experts on persistent organic pollutants. He and his colleagues tested the breast milk of Swedish women, and discovered the compounds there, too. He realized immediately what this meant: A new breed of contaminant, similar to PCBs, was rapidly spreading throughout the urban environment. It wouldn't be long before they permeated the Arctic, too. In the mid-1990s, the flame retardants, polybrominated diphenyl ethers, or PBDEs, were detected in Arctic wildlife. Today, they are found in every species tested, including polar bears, seals, beluga, killer whales, and seabirds.

The buildup of chemicals that Rachel Carson in 1962 dubbed the "chain of evil" continues, unbroken, to this day in the Arctic. Although old legacy contaminants such as PCBs and DDT are restricted or banned in many nations and declining in most Arctic regions, others are still infiltrating the far North and creating new toxic legacies. An array of flame retardants, pesticides, and other neurotoxins and hormone disruptors widely used in households, farms, and industries have taken their place, some of them building up in animal and human tissues at an extraordinary pace. "The past five years have seen a major increase in the number of halogenated organic chemicals detected in Arctic air," according to an AMAP report on POPS published in 2004.

PBDEs are the most worrisome of the new contaminants because they are increasing at a rate unseen since the PCBs and DDT of the 1960s. They were developed by chemical companies in the 1970s to protect the public by slowing the ignition and spread of fires in polystyrene foam and hard plastics that are used in upholstered furniture, textiles, carpet paddings, televisions, computers, and other electronic equipment. So far, the chemicals are in low concentrations in the Arctic—much lower than the levels found in animals and people in urbanized areas of the United States and Canada. Yet they are doubling approximately every seven years in many Arctic species. Michael Ikonomou, a scientist at Canada's Institute of Ocean Sciences, predicts that by 2050, brominated flame retardants will become the most prevalent compound in the seals of Arctic Canada. Already, in ringed seals, they increased ninefold between 1981 and 2000, while in beluga, they increased more than sixfold between 1982 and 1997. These flame retardants have already spread throughout the Arctic. Scientists reported in 2004 that ringed seals in remote regions of Russia, in the White, Barents and Kara seas, contain higher PBDE levels than seals in the Canadian Arctic. Because they are contaminating so many Inuit foods, including beluga and ringed seals, the flame retardants most likely have contaminated the Inuit and other Arctic people too, although as of early 2005, no human tests there had been completed.

Why do the tissues of animals so far from any city carry flame-retarding chemicals used in furniture and electronics? Scientists suspect that they are reaching the Arctic the same way that pesticides and PCBs have—by riding on winds and currents. They are slow to arrive in the higher latitudes, but since they are growing at astonishing rates, the peaks are yet to come. "In Arctic biota, the rapidly rising concentrations seen currently in Canada could be expected to continue for some time, reflecting continued production and use . . . in North America and the impact of long-range atmospheric transport," says a study published in 2003 by R. J. Law and other U.S. and Canadian scientists.

Today, PBDEs are more prevalent in urban areas than in the remote Arctic. The world's highest concentrations in wildlife, as of late 2004, are in fish-eating birds called Forster's terns nesting in San Francisco Bay. Throughout this decade and the last, the flame retardants have been growing exponentially in people and wildlife throughout the United States and Canada, doubling in concentration every few years. In half a century, since PCBs and DDT, no other toxic contaminant has spread as rapidly as they have. Ross Norstrom of the Canadian Wildlife Service, who has studied contaminants for thirty years, says he is "flabbergasted" by the rapid increases in North American animals—among them, a doubling in Great Lakes gulls, Lake Ontario trout, and San Francisco Bay seals every two to three years. The effects on wildlife are unknown, but some biologists suspect they could be having neurological and behavioral effects on animals that alter how they breed and raise their young.

What disturbs toxicologists the most about the new contaminants are their striking similarities to the old ones. PBDEs persist in the environment, accumulate in human and animal fat, and magnify up the food web just like PCBs. They are structurally similar in chemical composition to PCBs, and they have the same neurotoxic effects on the brains of newborn animals, at similar doses. Like the organochlorines of the past, the flame retardants pass through a mother's womb and are readily absorbed by a fetus.

Children of the 1960s and 1970s were heavily exposed to PCBs and DDT, while children of the 1990s—and beyond—are absorbing the new compounds along with residue of the old ones. "This will be a social experiment we'll be following for the next twenty years," says Kim Hooper of the California Environmental Protection Agency's Hazardous Materials Laboratory. "It is not going away."

No one knows for sure how the flame retardants are getting into human bodies but evidence in 2004 suggested it is coming mostly from indoor air. PBDEs have been found in the dust of most households and offices. The degree of contamination varies widely. Some homes, for unknown reasons, contain fifty times more than the average home. The chemicals appear to be off-gasing from furniture, carpets, computers, and other electronics. They are also found in a variety of food, especially fish, so they are contaminating the outdoor environment, too, and people are exposed through their diet. Studies of supermarket food in Texas and California found traces of the chemicals in virtually every product, including infant soy formula. That discovery, in particular, surprised experts because most other contaminants build up in animal fat, not in vegetables.

The effects of the new contaminants are eerily familiar. A single, relatively small dose of the flame retardants given to a newborn mouse or rat depletes thyroid hormones and consequently disrupts its developing brain, causing measurable changes in its learning ability, memory, behavior, and hearing, according to studies performed by three scientific teams. Ten-day-old mice fed PBDEs performed poorly in maze tests, were hyperactive, and were slower to habituate to new environments. The dose was fairly small, amounting to 4 parts per billion in the brain tissue of mice—only ten to one hundred times more than amounts found in some U.S. women and babies. "There is a window of (brain) development when these compounds cause effects, and these effects are persistent and worsen with age," says Swedish neurotoxicologist Per Eriksson of Uppsala University's Department of Environmental Toxicology. "PBDEs are as potent as PCBs in inducing an effect. We're quite sure that they cause the same effects at the same levels as PCBs."

Some Americans—perhaps 15 million people, according to one estimate—already have such high concentrations of these hormone-altering flame retardants in their bodies that it is possible that their children's brain development has been disrupted, scientists say. In the United States, tests on breast milk show women on average carry up to seventy times more PBDEs than women in Europe and Japan. Even the lowest amount found in American breast milk, 6 parts per billion, is more than twice Europe's average. In 2003, a pregnant Indiana woman had the largest individual concentration found, and her baby carried nearly as much at birth. A San Francisco Bay Area woman in her thirties was close behind. For PBDEs, "we're the third-world country," Hooper says. "Scientists looking for high exposure come here, to the U.S."

Swedish scientists who compared newly collected and archived samples of women's breast milk were shocked to find that women in the 1990s had 5,600 percent more PBDEs in their milk than women in the 1970s. The growth rate, says Stockholm University researcher Daiva Meironyte-Guvenius, "was very scary. The reaction here in Sweden was very powerful. We knew that if we were to continue that way, we could start to see effects in humans."

Stunned by the breast milk findings, published in 1998, electronics companies in Europe voluntarily phased out the octa and penta PBDEs used mostly in furniture foam and textiles, and as a result, concentrations in European breast milk dropped within a few years. California acted in 2003; its ban on the two compounds, octa and penta, will become effective in 2008. Since then, other states have followed suit. Facing pressure from the EPA and states, the U.S. manufacturer of penta and octa PBDEs discontinued their production at the end of 2004. Manufacturers are developing safer flame retardants for furniture foam, and one major furniture company, IKEA, sells only PBDE-free furniture.

Nevertheless, other PBDEs, particularly the most widely used one, called deca, found in the hard plastic of televisions, computers, and other electronics, remain legal. Scientists are finding low concentrations of deca throughout the world in water, air, wildlife, human blood, and women's breast milk. About 56,000 tons are used annually, mostly to make electronics equipment fire-resistant. Many experts worry that the bans on penta and octa might do little to reduce the contamination found in human bodies and the environment unless deca is banned, too. So far, there have been no government initiatives that phase out deca's use. The companies that manufacture PBDEs say they save lives by stopping flames from spreading quickly in electronics equipment. But many consumer electronics companies, including Intel, Toshiba, and Sony, are eliminating the chemicals, redesigning their products so that flame retardants aren't necessary.

PBDEs aren't the only "new" chemicals that are spreading globally and migrating to the Arctic. A perfluorinated compound called PFOS, perfluorooctane sulfonate, used in Scotchgard, the fabric protector, was found in Canadian polar bears in higher concentrations than many of the old compounds such as PCBs, according to a study published in 2004 by University of Toronto and Environment Canada scientists. Ringed seals—a species important to many Arctic predators—carry high levels of them, particularly in the southern parts of the Arctic. 3M Co. voluntarily discontinued use of the compound in 2002, but it is a persistent substance that remains in the environment, biomagnifying in the Arctic's marine food web. Related perfluorinated compounds, including one used in making Teflon, remain in use and have also been found in polar bears, seabirds, fish, foxes, seals, and other Arctic animals. So far, the perfluorinated chemicals are in higher concentrations in lower latitudes, but Derek Muir has found that, like PBDEs, they are increasing rapidly in polar bears and ringed seals. They likely are contaminating the Arctic's human inhabitants, too.

Other popular toxic compounds still in use, especially the insecticides endosulfan and methoxychlor, have also made their way to the Arctic. In humans and animals in the far North as well as in southern latitudes, the new contaminants are joining the old ones in tissues and blood, turning bodies into living repositories for toxic waste.

What will it take to prompt action to protect the world from contaminants spreading from the Arctic to Antarctica? Sweden's Åke Bergman says that time and time again, it has taken lethal or severe public health accidents to galvanize attention to the dangers. In the 1970s, polybrominated biphenyls (PBBs), fire-retarding chemicals similar to today's PBDEs, were outlawed in the United States only after they mistakenly were added to livestock feed and poisoned about 4,000 people in Michigan; PCBs were banned in much of the industrialized world after a similar accident in 1968 that sickened about 2,000 people in Japan who ate contaminated rice oil. People came down with skin diseases, reproductive disorders, and liver and stomach cancers. U.S. scientists knew less about the hazards of PBBs and PCBs when they were phased out in the 1970s than they now know about today's flame retardants, but for a variety of reasons—political, scientific, and legal—it is harder than ever to ban a chemical in the United States.

Bergman realized back in the 1980s that PBDEs posed a global threat but, like the PCBs and similar chemicals before them, they got little public attention until they showed up in tests of human breast milk, first in Sweden, then elsewhere. They are now, he says, the most troubling toxic contaminant in use today, spreading rapidly from pole to pole. "With the type of chemicals that PBDEs represent, it was quite obvious to me at an early point that this was going to be a new problem of major concern if nothing was done," Bergman says. "These are compounds that have the same properties as PCBs and DDT, and it's just a matter of concentration before we have a toxic effect. We already have fairly high concentrations in wildlife and in humans."

Scientists are dismayed that society has learned nothing from the toxic legacies of the past and appears destined to repeat them over and over. Only the names of the chemicals change.

"Those of us who have been around for quite a few years, we ask, 'didn't we learn anything from PCBs?' In my darker moments," Bergman says, "I wonder—do we need another accident to get people to act?"

Epilogue

Survival of the Fittest:
Walking in the Inuit's Footsteps

Once upon a time, in an old Inuit legend, the hunter Kiviuq met a beautiful woman who in reality was a goose. He married her anyway and as time passed, the goose-woman decided she wanted to eat her own foods instead of the caribou and seal meat that her husband liked. Kiviuq insisted that she eat human food, so the goose-woman gathered her children and flew with them far away, to the south. When he returned, his family was gone. He searched for them everywhere. One day he met a man who created fish from wood. He made Kiviuq a large fish to carry him to sea to search for his family. He found them and decided that he didn't care what they ate. They returned north, together, and each let the other live as he or she wished.

Comparable to Homer's *Odysseus,* Kiviuq is a wandering hero, a hunter who encounters ever-present danger, managing to survive harsh, unpredictable conditions in the Arctic with ingenuity and patience and good luck. Thus, says a sign at an exhibit at the Musée d'Art Inuit Brousseau, Kiviuq's tale is a "metaphor for Inuit existence."

I am visiting the museum in Québec City after leaving Eric Dewailly's office at Laval University's public health department. It is late September but summer lingers. Autumn leaves have not yet begun to fall and this historic city, fortified by old stone walls, seems suspended in time. Québec, more European than North American, seems to have a split personality, undecided over its identity. The leaders of this province are just as undecided about how to treat aboriginal people, the inhabitants of Nunavik, Québec's far north. I left Dewailly's office a bit unnerved by our discussion about his commitment to do no harm, wondering whether Canadians were doing enough to protect the Inuit from contaminants in their food. I craved an instant infusion of Inuit advice and headed to the museum. Its galleries display all forms of Canadian Inuit art, but my personal favorites are the miniatures from Nunavik, the finely crafted ivory sculptures of seal, walrus, seabirds, and other animals that remind me of Native American Zuni fetishes. At the museum, a sign over the artistic depictions of their prey grabs my eye: "Hunting is no longer vital to Inuit survival, yet it remains an important aspect of the Inuit economy and lifestyle, and even Inuit identity." No longer vital to survival? I'm puzzled by that assertion, wondering how the museum curators would define those words. In my journeys, the most enduring lesson I learned is that hunting is as vital to Arctic people as it was centuries ago.

It is obvious even today, in this time of political correctness toward most native peoples, that the Arctic way of life is still misunderstood. As I left Québec, I found myself hoping that Canadians, the only ones who have been bold and brave enough to address the human contamination dilemma head-on in the Arctic, will continue to have the courage to keep searching for answers, even if they make more mistakes along the way. I think about what Jim Estes, a scientist for almost forty years who has watched ecosystems fall apart and species vanish before his very eyes, told me during one of our many hourslong conversations about the nature of uncertainty and the uncertainty of nature. How should society react in the face of so many unknowns? Break the paralysis, he insists. Act on consensus, not just certainty. Treat the environment like it's a civil lawsuit—a preponderance of evidence is enough to convict a bad actor—rather than prove guilt beyond a reasonable doubt. He is frustrated, and a bit distraught, that years—decades—of research like his own have accomplished little that actually protects the resources. I personally do not believe that the end always justifies the means when it comes to environmental protection. I do not agree that any movement is better than no movement at all, that the precautionary principle must guide everything we do. Sometimes the cost to society (and I don't mean just dollars and cents) is too high, the benefits too scant. But in the case of Arctic inhabitants, our unintentional lab rats, I do know that scientists and policy makers cannot dare to wait for certainty. Clarity should be enough to write a prescription for the Arctic, for we are absolutely certain of one thing: The toxic contaminants in their foods are not good for them—or for us.

Québec was the final journey of my project, and, like the scientists there, I didn't find all the answers I had hoped for. But my quest continues, in the Arctic and in new venues. I returned to the *Los Angeles Times* with even more of a compulsion to understand the effects of chemicals on people and wildlife, not just the relics of the past but newer compounds found in many household products today. The legacy contaminants have declined substantially throughout most of the world since I was a child in the 1960s. For a time, I felt reassured that the worst was over, that my own child would not have to live with an inordinate burden of chemicals building up in his own body and in the bodies of creatures around him. But now, with the discovery that brominated flame retardants, PBDEs, are ubiquitous, spreading globally and rapidly accumulating in people and wildlife just like the old PCBs, I realize that we haven't cured the problem, we have simply moved it into a new phase. Our parents bequeathed to us one toxic threat; we have bequeathed to our children another. After a massive, multimillion-dollar cleanup, Lake Michigan is no longer the nation's PCB hot spot like it was when I was growing up along its shore, but now its fish carry some of the highest concentrations of PBDEs found in fish anywhere in the world, and the levels are still growing. Our generation simply exchanged an old threat for a new one. We have left our children and grandchildren—as well as inhabitants of the faraway, frozen North—a new chemical inheritance. This is a perverse variation of Earth's circle of life, and I can't help but wonder if it is adding a new twist to Charles Darwin's theory of natural selection, the survival of the fittest. Perhaps the creatures that ultimately will survive are those able to cope with man-made chemicals scrambling their hormones, disarming their immune systems, and tampering with the inner workings of their brains.

No one is certain what, if any, impacts this brew of toxic substances is having on human and wildlife health, but knowledge is power, so we must do everything we can to find out. We can all be relieved that scientists, so far, have uncovered no dead bodies, no obvious, noticeable health disorders in the people and animals of the Arctic tied to contaminants. Children of the Arctic aren't dying from these chemicals. They aren't left retarded or infertile. But they may not be reaching their full cognitive potential and they may be sick from infections, and that would be a tragic enough legacy for the world to impose on them. Every child deserves a chance to develop normally without a chemical stigma. The Inuit way of life is important to maintain, and so is the nutrition provided by traditional foods, but I also am left with the feeling that no food is worth endangering a child for.

All of us—the *Qallunaat*, the people of the south—have a chance to do something now to guide our future actions and fix our mistakes of the past by vowing to never repeat them. If we turn our backs on the Arctic's plight—out of sight, out of mind—we will continue to endanger the cultures and ecosystems of distant lands far into the future. The governments and corporations of the industrialized world must make efforts to understand the global impacts of all the chemicals they use, and restrict or replace those that are becoming globetrotters, spreading and building up and persisting in the environment for decades. They must also strive to eliminate or contain the huge dumpsites and stockpiles of obsolete compounds left behind. The best environmental solutions come from actions taken by consumers as well as industries and government. Still, there is little that we as consumers can do, except to ask more questions about the products we buy and the chemicals we use.

Stopping all global contaminants is impractical and unenforceable, and I don't have to be a prophet to predict that some will continue to build up in the Arctic, and many new ones will be discovered there. Because of this reality, leaders of Arctic nations and indigenous groups must face the fact that chemicals will always trespass in their homeland so they need to find culturally sensitive ways to reduce consumption of the most tainted foods. Inuit in Greenland, in particular, as well as Russia and Canada's Nunavut and Nunavik, must take action as quickly as possible to reduce their exposure, without waiting for scientists to figure out more about how their health is being affected. Arctic hunters may have to forsake certain prey and rely on less-contaminated ones, preserving their health as well as their heritage.

The Arctic's inhabitants have endured for millennia because of their ability to evolve, their willingness to transform themselves to adapt to their hostile, ever-changing environment, whether natural or manmade. With little power to defend themselves, they are at the mercy of others—governments and outsiders who have long ignored their needs—or worse, tried to wipe out their ancient traditions by assimilating them into modern society. Arctic people have often been treated like foreigners in their own land, yet efforts to shatter their communities and cultures inevitably fail. Canada's Inuit retained their heritage despite the government's separating their children from their families and sending them away for schooling for more than a century. The people of the Thule region of Greenland remained together, creating the village of Qaanaaq, even when their entire community was evicted by the U.S. and Danish governments in the 1950s so that a military base could be established there during the Cold War. And Alaska's Inupiat are still avid hunters despite an infusion of oil money and continuing international efforts to ban their whaling. Nevertheless, the coming decade is a critical one. Never before has the Arctic had to weather so many concurrent pressures from outside forces—contaminants and climate change and modernization. Scientists say it is unlikely that Arctic cultures and ecosystems will escape intact.

The stresses of the Arctic are supposed to come from the harshness of nature, the brutality of its weather and terrain, not from the careless hand of man. After generations of honing their skills and tools, Arctic inhabitants know how to survive violent storms, months-long spans of darkness, and temperatures plunging to 60 below zero Fahrenheit. But there is nothing in their traditional knowledge, nothing they can borrow from the wisdom of elders or the defensive skills of wildlife or the nuances of nature, that will help them survive exposure to toxic chemicals.

Yet with patience and ingenuity and good luck, Arctic inhabitants —like the legendary Kiviuq—will survive, perhaps even thrive. Despite the harsh conditions they live in, they request nothing, no help from the outside world, except to be left alone. They know that near the North Pole, survival is something that is earned, never guaranteed. The Inuit and Inupiat believe that the power to influence their own destiny lies in their own hands, that they will continue to be bestowed with nature's hearty and healthy gifts as long as they treat the Earth and its creatures with respect. But they also have faith that if they become lost somewhere along the way, an Inuksuk, embodied in the traditions of their elders and forefathers, will guide them like a compass down the proper path. Their only regret is that the rest of the world doesn't follow their creed. "We as Inupiat have been the guardians of the Arctic for thousands of years," says Charlie Hopson, who has spent a half-century hunting bowhead in Alaska. Only death, he says, will stop him from pursuing that destiny. Feeling neither hopeless nor helpless, the people of the Arctic embrace their plight, patiently and intuitively awaiting their fate.

In Inuit legend, the goddess Sedna rewards the people of the land with the food of the sea. Once upon a time, when Sedna was young and beautiful, her father, a widower, forced her to marry his dog because she stubbornly wouldn't marry the suitor of his choice. She gave birth to children, some dogs, some human, and she set them all adrift in the

Arctic. The dogs were the ancestors of white men while the humans were the ancestors of the Inuit, and she sent them floating out to sea in different directions. Hungry, she married a hunter who turned out to be a fulmar, a seabird that lives on the open ocean. Her father attempted to take her away in a kayak, but the seabird tried to stop them, stirring up a violent storm. Fearing for his own life, the father threw his daughter overboard, and when she clung to the kayak, he chopped off her fingers, which fell into the sea. One finger was transformed into a seal, one a walrus, one a whale, one a polar bear, and all the marine mammals on Earth were imbued with the spirit of a human. Sedna was transformed into a compassionate yet strong-willed mother of the sea, a goddess who rules over all life and nourishes the bodies and souls of the Inuit. Whenever a sea animal dies, its soul returns to Sedna and reports if any of her laws are being broken by human beings. When Sedna is pleased with how her creatures are being treated, she shares her bounty with the Inuit. When they have angered her, they starve and fall ill. Over time, all of mankind's sins against nature fall down through the water, and the filth collects in Sedna's hair. A shaman, a human Inuk with great power, must help untangle her from the mess by descending into the sea, combing her hair, and confessing the people's sins, promising never to repeat them. Only then can the Inuit thrive again.

This spirit of the Arctic is captured in a token that Mamarut presented to me at the end of our narwhal-hunting journey in Qaanaaq. Carved from a reindeer antler by his ancestors, it is a finger-sized totem, an old, weathered sculpture of a human face. I now see it as a most fitting gift, since Mamarut and all the others I came to know put human faces on the Arctic Paradox, seeking to pass to me the eternal wisdom of their forefathers. In my home, I display it next to a miniature stone Inuksuk handcrafted in Nunavik. Every time I see these carvings, I am reminded of Mamarut's warning that if I am to stay safe on my ventures, I must walk only where he has left his footprints. The inhabitants of the Arctic are providing us an Inuksuk, showing us a safe passage as we struggle to make wise decisions in modern times. Tread in our footsteps, they say, for we know the way.

In the very earliest time,
when both people and animals lived on earth,
a person could become an animal if he wanted to and an animal could
become a human being.
Sometimes they were people
and sometimes animals
and there was no difference.
All spoke the same language.
That was the time when words were like magic.
The human mind had mysterious powers.
A word spoken by chance
might have strange consequences.
It would suddenly come alive
and what people wanted to happen could happen—
all you had to do was say it.
Nobody could explain this:
That's the way it was.

—Edward Field, *Magic Words*, 1998,
based on a centuries-old Inuit story

Acknowledgments

I never would have been able to pursue this project if it weren't for the Pew Fellows Program in Marine Conservation, which funded it in its entirety through a fellowship. I was nominated for the 1999 award by Jim Detjen, a fellow environmental journalist and educator, and when I was asked to conceive a project, I knew that no topic could compare with the critical need to explore the relentless contamination of the Arctic. Thankfully, the judges (journalists and scientists) agreed. Newspapers and other media outlets are often unable to devote the substantial resources necessary for intense travel and research in remote lands, so fellowships like this are invaluable to journalists covering science and the environment. I thank the Pew Fellows Program for trusting a journalist with an award that is usually reserved for scientists, and allowing me to pursue this research with no strings attached. Their goal, and mine, was simply to seek the truth about global contaminants in the marine environment.

Special thanks to the people who not only opened their minds to me, but sometimes their homes. Ingmar Egede, more than anyone else I encountered on my trips, helped me understand the Inuit's inexorable ties with nature. Ingmar generously shared his world, so unlike my own. He became my personal Inuksuk, moral and physical, in all my Arctic journeys. The indigenous people of the world are mourning the loss of Ingmar, one of the most eloquent promoters of their rights, who died in August 2003. May you rest in peace, Ingmar, in a place where the bounties of the sea are endless and the harmony between cultures eternal.

Even on bitterly cold days, I felt warmth emanating from the Arctic people, an inner warmth that you can find only in distant, harsh lands where people must rely on each other. I am especially grateful to Ólavur Sjurdarberg for his hospitality during a day that was difficult for all Americans. I was alone, traveling by bus to Olavur's tiny village in the Faroe Islands, unable to comprehend a single word of Faroese, when I kept hearing the English words "World Trade Center" repeated over and over on a blaring radio. I stepped off the bus to hear Olavur tell me in broken English that the New York towers had collapsed and several jets had been hijacked. I was probably the only American in the Faroe Islands on September 11, 2001, and one of the last to learn of the events that day, but Olavur made sure I didn't feel alone, offering me some delicious cod he had just caught, and friendship that transcends all borders.

I am especially grateful to the dozens of international scientists, particularly from Canada, Denmark, and Norway, associated with the Arctic Monitoring and Assessment Programme (AMAP) who trusted me with their knowledge of Arctic biology, chemistry, and sociology, and helped me understand places and phenomena that were, at first, so foreign to me. I am especially grateful to Derek Muir, Pál Weihe, Philippe Grandjean, Eric Dewailly, Andy Derocher, Rob Macdonald, Jim Estes, Donald Mackay, Lars-Otto Reiersen, and Christopher Furgal. Thank you for your confidence and patience.

With the enthusiasm of literary agent Russell Galen and the intuitive editing skills of Grove Atlantic's Brando Skyhorse, I felt empowered to fashion a story that could resonate far beyond the Arctic Circle. Special thanks to Steve Gugerty for gracing these pages with his artwork. Thanks also to the *Los Angeles Times* for giving me the time off to pursue this project, and to my friends who gave me the confidence to do it and cajoled me to finish it.

I am especially grateful to my parents, who never once doubted me in whatever I ventured to do, nor tried to tell me that writing was no way to make a living. During the fog of research, my son's kindergarten class at Lowell Elementary School in Long Beach, California—Linda Nyquist's class of 2002–2003—inspired me to remember what was at stake with future generations. And finally, I thank my husband, Dan, and my son, Christopher, who sacrificed a lot during the years it took for me to tell this tale.

Bibliography

Addison, R. F., and P. F. Brodie. Organochlonne residues in maternal blubber, milk, and pup blubber from grey seals (*Halichoerus grypus*) from Sable Island, Nova Scotia. *J Fish Res Board Can* 1977, 34:937–41.

Addison, R. F., and T. G. Smith. Organochlorine residue levels in Arctic ringed seals: Variation with age and sex. *Oikos* 1974, 25(3):335–77.

Alaska Geographic Society. *The Aleutian Islands.* Alaska Geographic Society, 1995.

Alexander, Bryan, and Cherry Alexander. *The Vanishing Arctic.* Facts on File, 1997.

Anthony, R., et al. Productivity, diets and environmental contaminants in nesting bald eagles from the Aleutian Archipelago, Alaska. Abstracts, Society of Environmental Toxicology and Chemistry Annual Meeting, 2004.

Arctic Climate Impact Assessment. Cambridge University Press, 2004.

Arctic Climate Impact Assessment. Overview Report. Fourth Arctic Council Ministerial Meeting, November 2004.

Arctic Monitoring and Assessment Programme. *AMAP Assessment Reports.* 1998 and 2002.

———. Impacts of POPs and mercury on Arctic environments and humans. AMAP conference and workshop. Abstracts, 2002.

———. The influence of global change on contaminant pathways to, within, and from the Arctic. 2003.

———. Persistent toxic substances, food security and indigenous peoples of the Russian North. AMAP, UNEP, and Russian Association of the Indigenous Peoples of the North, 2004.

Ayotte, P., et al. PCBs and dioxin-like compounds in plasma of adult Inuit living in Nunavik (Arctic Québec). *Chemosphere* 1997, 34(5–7):1459–68.

Bernhoft, A., et al. Organochlorines in polar bears (*Ursus maritimus*) at Svalbard. *Environ Pollut* 1997, 95:159–75.

———. Possible immunotoxic effects of organochiorines in polar bears (*Ursus maritimus*) at Svalbard. *J Toxicol Environ Health A* 2000, 59(7): 561–74.

Berton, Pierre. *The Arctic Grail: The Quest for the Northwest Passage and the North Pole, 1818–1909*. Lyons Press, 2000.

Bjerregaard, P., et al. Exposure of Inuit in Greenland to organochlorines through the marine diet. *J Toxicol Environ Health A* 2001, 62(2):69–81.

Bockstoce, John R. *Whales, Ice, and Men: The History of Whaling in the Western Arctic*. University of Washington Press, 1989.

Braathen, M., et al. Relationships between PCBs and thyroid hormones and retinol in female and male polar bears. *Environ Health Perspect* 2004, 112(8):826–33.

Bratton, G., et al. Assessment of selected heavy metals in liver, kidney, muscle, blubber and visceral fat of eskimo harvested bowhead whales from Alaska's North Coast. Alaska Department of Wildlife Management, 1997.

Braune, B., et al. Changes in concentrations of organochlorines and mercury in Canadian Arctic seabirds between 1975 and 2003. Abstracts, Society of Environmental Toxicology and Chemistry, 2004.

Buell, Janet. *Greenland Mummies*. Twenty-first Century Books, 1998.

Carlsen, E., et al. Evidence for decreasing quality of semen during past fifty years. *BMJ* 1992, 305(6854):609-13 (review).

Carson, Rachel. *Silent Spring*. Houghton Mifflin Company, 1962.

Chabot, Marcelle. As long as I am not hungry: Socio-economic status and food security of low-income households in Kuujjuaq. Nunavik Regional Board of Health and Social Services, 2004.

Colborn, Theo, et al. Developmental effects of endocrine-disrupting chemicals in wildlife and humans. *Environ Health Perspect* 1993, 101:378–84.

———. *Our Stolen Future: Are We Threatening Our Fertility, Intelligence, and Survival? A Scientific Detective Story*. E. P. Dutton, 1996.

Crawford, Michael, and David Marsh. *The Driving Force: Food, Evolution, and the Future*. Harper & Row, 1989.

Dallaire, F., et al. Acute infections and environmental exposure to organochlorines in Inuit infants from Nunavik. *Environ Health Perspect* 2004, 112(14):1359–65.

Darwin, Charles. *On the Origin of Species*. John Murray, 1859.

Derocher, Andrew, et al. Contaminants in Svalbard polar bear samples archived since 1967 and possible population level effects. *Sci Total Environ* 2003, 301(1–3):163–74.

De Swart, R. L., et al. Impaired immunity in harbour seals (*Phoca vitulina*) fed environmentally contaminated herring. *Vet Q* 1996, 18(Suppl. 3):S127–8 (review).

Dewailly, Eric, et al. Concentration of organochlorines in human brain, liver, and adipose tissue autopsy samples from Greenland. *Environ Health Perspect* 1999, 107(10):823–8.

———. High levels of PCBs in breast milk of Inuit women from Arctic Québec. *Bull Environ Contam Toxicol* 1989, 43(5):641–6.

———. Inuit are protected against prostate cancer. *Cancer Epidemiol Biomarkers Prev* 2003, 12(9):926–7.

———. Inuit exposure to organochlorines through the aquatic food chain in Arctic Québec. *Environ Health Perspect* 1993, 101(7):618–20.

———. Susceptibility to infections and immune status in Inuit infants exposed to organochlorines. *Environ Health Perspect* 2000, 108(3):205–11.

Donne, John. *The Complete Poetry and Selected Prose of John Donne*. Modern Library, 2001.

Downie, David Leonard, and Terry Fenge. *Northern Lights Against POPs*. McGill-Queen's University Press, 2003.

Dupre, Lonnie. *Greenland Expedition: Where Ice Is Born*. Northword Press, 2000.

Egede, Ingmar. Inuit food and Inuit health: Contaminants in perspective. Inuit Circumpolar Conference, Seventh General Assembly, 1995.

Estes, James, et al. Killer whale predation on sea otters linking oceanic and nearshore ecosystems. *Science* 1998, 282(5388):473–6.

———. Organochlorines in sea otters and bald eagles from Aleutian Archipelago. *Mar Pollut Bull* 1997, 34(6):486–90.

Field, Edward, and Stefano Vitale. *Magic Words*. Gulliver Books, 1998.

Fisk, A., et al. Metals in apex predators of the Arctic. Abstracts, Society of Environmental Toxicology and Chemistry Annual Meeting, 2004.

Freeman, Milton M. R., et al. *Inuit, Whaling and Sustainability*. AltaMira Press, 1998.

Fry, D. M. Reproductive effects in birds exposed to pesticides and industrial chemicals. *Environ Health Perspect* 1995, 103(Suppl. 7):165–71.

Fry, D. M., et al. DDT-induced feminization of gull embryos. *Science* 1981, 213(4510):922–4.

Grandjean, Philippe, et al. Cardiac autonomic activity in methylmercury neurotoxicity: Fourteen-year follow-up of a Faroese birth cohort. *J Pediatr* 2004, 144(2):169–76.

———. Cognitive deficit in seven-year-old children with prenatal exposure to methylmercury. *Neurotoxicol Teratol* 1997, 19(6):417–28.

———. Cognitive performance of children prenatally exposed to "safe" levels of methylmercury. *Environ Res* 1998, 77(2):165–72.

———. Neurobehavioral deficits associated with PCB in seven-year-old children prenatally exposed to seafood neurotoxicants. *Neurotoxicol Teratol* 2001, 23(4):305–17.

Greenland in Figures 2003. Greenland Home Rule Government.

Guillette, Louis, et al. Developmental abnormalities of the gonad and abnormal sex hormone concentrations in juvenile alligators from contaminated and control lakes in Florida. *Environ Health Perspect* 1994, 102(8):680–8.

Haave, M., et al. Polychlorinated biphenyls and reproductive hormones in female polar bears at Svalbard. *Environ Health Perspect* 2003, 111(4):431–6.

Hansen, Kjeld. *A Farewell to Greenland's Wildlife*. Gads Forlag, 2002.

Hess, Bill. *Gift of the Whale: The Inupiat Bowhead Hunt, a Sacred Tradition.* Sasquatch Books, 2003.

Hightower, Jane, et al. Mercury levels in high-end consumers of fish. *Environ Health Perspect* 2003, 111(4):604–8.

Hisdal, Vidar. *Svalbard: Nature and History*. Norsk Polarinstitutt, 1998.

Hoekstra, P. F., et al. Bioaccumulation of organochlorine contaminants in bowhead whales (*Balaena mysticetus*) from Barrow, Alaska. *Arch Environ Contam Toxicol* 2002, 42(4):497–507.

Holden, A. V. Monitoring organochlorine contamination of the marine environment by the analysis of residues in seals. Paper presented at FAO Conference on Marine Pollution, 1970.

———. Organochlorine residues in seals. Paper presented at the International Council for the Exploration of the Sea, 1969.

———. Source of polychlorinated biphenyl contamination in the marine environment. *Nature* 1970, 228(277):1220–1.

Holden, A. V., et al. Organochlorine pesticides in seals and porpoises. *Nature* 1967, 216(122):1274–6.

Ikonomou, M. G., et al. Exponential increases of the brominated flame retardants, polybrominated diphenyl ethers, in the Canadian Arctic from 1981 to 2000. *Environ Sci Technol* 2002, 36(9):1886–92.

Intergovernmental Panel on Climate Change (IPCC). Third Assessment Report, Climate Change 2001.

International Labor Organization. *Traditional Occupations of Indigenous and Tribal Peoples: Emerging Trends.* Cnossos/Dumas-Titoulet Imprimeurs, 2000.

Kinloch, D., and H. V. Kuhnlein. Assessment of PCBs in Arctic foods and diets. A pilot study in Broughton Island, Northwest Territories, Canada. *Arctic Med Res* 1988, 47(Suppl. 1):159–62.

———. Inuit foods and diet: A preliminary assessment of benefits and risks. *Sci Total Environ* 1992, 122(1–2):247–78.

Kuhnlein, H. V., et al. *Assessment of Dietary Benefit/Risk in Inuit Communities.* Centre for Indigenous Peoples' Nutrition and Environment, McGill University, 2000.

Law, R. J., et al. Levels and trends of polybrominated diphenylethers and other brominated flame retardants in wildlife. *Environ Int* 2003, 29(6):757–70 (review).

Li, Y. F., et al. Historical alpha-HCH budget in the Arctic Ocean: The Arctic Mass Balance Box Model (AMBBM). *Sci Total Environ* 2004, 324 (1–3):115–39.

Lie, E., et al. Does high organochlorine (OC) exposure impair the resistance to infection in polar bears (*Ursus maritimus*)? Part I: Effect of OCs on the humoral immunity. *J Toxicol Environ Health A* 2004, 67(7):555–82.

Lindberg, S. E., et al. Dynamic oxidation of gaseous mercury in the Arctic troposphere at polar sunrise. *Environ Sci Technol* 2002, 36(6):1245–56.

Lopez, Barry. *Arctic Dreams.* Charles Scribner's Sons, 1986.

Lowenstein, Tom. *Ancient Land: Sacred Whale.* Harvill Press, 1999.

Lundgren, Stefan, and Olle Carlsson. *Svalbard: The Land Beyond the Northcape.* Ice Is Nice Publishing House, 1999.

Lyketsos, G. Should pregnant women avoid eating fish? Lessons from the Seychelles. *Lancet* 2003, 361(9370):1667–8.

McGrath-Hanna, Nancy K., et al. Diet and mental health in the Arctic: Is diet an important risk factor for mental health in circumpolar peoples? A review. *Int J Circumpolar Health* 2003, 62:3.

Muckle, Gina, et al. Prenatal exposure of Canadian children to polychlorinated biphenyls and mercury. *Can J Public Health* 1998, 89(Suppl. 1): S20–5, 22–7.

Muir, Derek, et al. Arctic marine ecosystem contamination. *Sci Total Environ* 1992, 22(1–2):75–134 (review).

———. Spatial and temporal trends and effects of contaminants in the Canadian Arctic marine ecosystem: A review. *Sci Total Environ* 1999, 230 (1–3):83–144 (review).

Mulvad, Gert, and Henning Sloth Pedersen. Benefit and risk of traditional food for indigenous Peoples. 11th Inuit Studies Conference, 1998.

Murata, K., et al. Delayed brainstem auditory evoked potential latencies in fourteen-year-old children exposed to methylmercury. *J Pediatr* 2004, 144(2): 177–83.

———. Evoked potentials in Faroese children prenatally exposed to methylmercury. *Neurotoxicol Teratol* 1999, 21(4):471–2.

Myers, G. J., et al. Prenatal methylmercury exposure from ocean fish consumption in the Seychelles child development study. *Lancet* 2003, 361(9370):1686–92.

Nansen, Fridtjof. *Farthest North.* Modern Library, 1999.

National Research Council. *Climate Change Science: An Analysis of Some Key Questions.* National Academies Press, 2001.

———. *The Decline of the Steller Sea Lion in Alaskan Waters: Untangling Food Webs and Fishing Nets.* National Academies Press, 2003.

———. *Toxicological Effects of Methylmercury.* National Academies Press, 2000.

Northern Contaminants Program, Minister of Indian Affairs and Northern Development. *Canadian Arctic Contaminants Assessment Reports,* 1997 and 2003.

O'Hara, T. M., et al. Organochlorine contaminant levels in Eskimo harvested bowhead whales of Arctic Alaska. *J Wildl Dis* 1999, 35(4):741–52.

Oskam, I. Organochlorines affect the major androgenic hormone, testosterone, in male polar bears (*Ursus maritimus*) at Svalbard. *J Toxicol Environ Health A* 2003, 66(22):2119–39.

———. Organochlorines affect the steroid hormone cortisol in free-ranging polar bears (*Ursus martimus*) at Svalbard, Norway. *J Toxicol Environ Health A* 2004, 67(12):959–77.

Peary, Robert E. *The North Pole*. Cooper Square Press, 2001.

Persistent Toxic Substances, Food Security and Indigenous Peoples of the Russian North. AMAP, 2004.

Petursdottir, Gudrun. *Whaling in the North Atlantic*. Fisheries Research Institute, University of Iceland, 1997.

Philip, Neil. *Songs Are Thoughts: Poems of the Inuit*. Orchard Books, 1995.

Rasmussen, Knud. *Across Arctic America: Narrative of the Fifth Thule Expedition*. University of Alaska Press, 1999.

Rice, D., et al. Critical periods of vulnerability for the developing nervous system: Evidence from humans and animal models. *Environ Health Perspect* 2000, 108(Suppl. 3):511–33 (review).

Rink, Hinrich. *Tales and Traditions of the Eskimo*. Dover Publications, 1997.

Ross, Peter. Endangered NE Pacific southern resident killer whales are at risk for contaminant-related health impacts. Abstracts, Society of Environmental Toxicology and Chemistry Annual Meeting, 2004.

———. Marine mammals as sentinels in ecological risk assessment. *Human and Ecological Risk Assessment* 2000, 6:29–46.

———. *Seals, Pollution and Disease: Environmental Contaminant-Induced Immunosuppression*. Universiteit Utrecht, 1995.

Ross, Peter, et al. High PCB concentrations in free-ranging Pacific killer whales, *Orcinus orca:* Effects of age, sex and dietary preference. *Marine Poll Bull* 40:504–15.

Savinova, T., et al. Persistent organic pollutants in ringed seals from the White, Barents and Kara Seas, Russia. Abstracts, Society of Environmental Toxicology and Chemistry Annual Meeting, 2004.

Schroeder, William, et al. Arctic springtime depletion of mercury. *Nature* 1998, 394:331–2.

Searles, Edmund. The politics and symbolics of seal hunting and seal meat: Case studies from South Baffin (Nunavut). Laval University. Abstract, 2004, International Arctic Social Sciences Association.

Seidelman, Harold, and James Turner. *The Inuit Imagination: Arctic Myth and Sculpture*. University of Alaska Press, 2001.

Skaare, J. U., et al. Organochlorines in top predators at Svalbard: Occurrence, levels and effects. *Toxicol Lett* 2000, 112–13:103–9.

————. Relationships between plasma levels of organochlorines, retinol and thyroid hormones from polar bears (*Ursus maritimus*) at Svalbard. *J Toxicol Environ Health A* 2001, 62(4):227–41.

Sonne, C., et al. Enlarged clitoris in wild polar bears (*Ursus maritimus*) can be misdiagnosed as pseudohermaphroditism. *Sci Total Environ* 2005, 337(1–3):45–58.

————. Is bone mineral composition disrupted by organochlorines in east Greenland polar bears (*Ursus maritimus*)? *Environ Health Perspect* 2004, 112(17):1711–6.

Springer, A. M., et al. Sequential megafaunal collapse in the North Pacific Ocean: An ongoing legacy of industrial whaling? *Proc Natl Acad Sci* 2003, 100(21):12,223–8.

Steuerwald, U., et al. Maternal seafood diet, methylmercury exposure, and neonatal neurologic function. *J Pediatr* 2000, 136(5):599–605.

United Nations Environment Programme. Arctic Environment: European Perspectives, 2004.

————. Global Mercury Assessment: Report of the Executive Director and Report of the Governing Council, Twenty-second session, 2003.

Usher, Peter, et al. *Communicating About Contaminants in Country Food: The Experience in Aboriginal Communities.* Research Department, Inuit Tapirisat of Canada, 1995.

Vaughan, Richard. *The Arctic: A History.* Alan Sutton Publishing Limited, 1994.

Von Finckenstein, Mary, et al. *Celebrating Inuit Art, 1948–1970.* Key Porter Books, 2000.

Wallace, Mary. *The Inuksuk Book.* Maple Tree Press, 1999.

Wania, F., and D. Mackay. Tracking the distribution of persistent organic pollutants. *Env Sc Techn* 1996, 30(9):390A–96A.

Weihe, Pál, et al. Neurobehavioral performance of Inuit children with increased prenatal exposure to methylmercury. *Int J Circumpolar Health* 2002, 61(1):41–9.

Wiig, O., et al. Female pseudohermaphrodite polar bears at Svalbard. *J Wildl Dis* 1998, 34(4):792–6.

Wolfson, Evelyn. *Inuit Mythology.* Enslow Publishers, 2001.

Index